# THE BOOK OF PANDAS, KANGAROOS, AND SNOW MONKEYS

# THE BOOK OF PANDAS, KANGAROOS, AND SNOW MONKEYS

*Margaret Rau*

AN AUTHORS GUILD BACKINPRINT.COM EDITION

*The Book of Pandas, Kangaroos, and Snow Monkeys*

All Rights Reserved © 1979, 2001 by Margaret Rau

AN AUTHORS GUILD BACKINPRINT.COM EDITION

Published by iUniverse.com, Inc.

For information address:
iUniverse.com, Inc.
5220 S 16th, Ste. 200
Lincoln, NE 68512
www.iuniverse.com

Originally published by Knopf

ISBN: 0-595-19826-0

Printed in the United States of America

# Foreword

To MANY PEOPLE, especially those who have seen the giant panda in zoos around the world, the animal seems like a lovable toy come to life, an animated childhood fantasy. Its many near-human characteristics and its sad clown face have fostered an all too anthropomorphic interpretation of its behavior.

It was my wish in this book to remove the panda from the artificial atmosphere of the zoo and return it to its native habitat, the mountain wilderness of Szechuan Province in far western China. My aim in doing this was to bring out the animal's basic nature as it is expressed in the wild.

Since no creature can exist by itself, I have tried as well to re-create a total picture of the wilderness through which the panda roams. This includes descriptions of those other rare creatures whose habitat it shares.

Much of my information about the panda comes from the 1974 *Acta Zoologica Sinica*, the journal of the Peking Zoological Gardens, which I obtained during a recent visit to the Peking zoo and subsequently translated. The information in this journal includes not only the care, feeding, and breeding of zoo pandas, but also the details of several expeditions to Szechuan's Wanglang Preserve to observe the animal in its native home.

I have also had recourse to English translations of the writings of Ernst Schaefer, a German naturalist who in the 1930s spent some time studying the fauna and flora of the region. And I have read voluminously from other explorers, naturalists, and hunters who in previous years ventured into the mountainous areas of far western China. For those who are interested in more complete scientific and historical evidence available on the giant panda, I have included in the Epilogue additional technical information about the panda's evolution and details about the history of man's encounters with—and effect on—the panda in its natural habitat.

Using all sources available I have attempted to depict as faithfully as possible the varied activities and behavior of the giant panda at home. It is my hope that this book may serve as a tribute to this strange and wonderful animal.

# The Giant Panda at Home

*Panda Country*

# Prologue:
# The Home of
# the Giant Panda

THE HOME of the giant panda lies in some of the wildest country in the world—the towering mountainous region of far western China. Here above the rich, low-lying farmlands of the Szechuan basin rises a broken wilderness cut by deep ravines and almost perpendicular slopes. Ridge after ridge mounts upward to the highest peaks, some 24,000 feet above sea level, brooding beneath their glistening white caps of perpetual snow.

Heavy rainfall has made the ravines and lower slopes of the mountains lush with greenery. They are crisscrossed with streams and rivulets that plunge down fiercely in foaming cataracts and waterfalls flashing with spray. A broad forest belt circles the mountains. In it are many kinds of deciduous trees, which lose their leaves in winter—oaks and red birches, poplar and chestnut. Mixed in among them are the evergreen trees, magnolia and hemlock, spruce and fir and pine. Here grows the dawn redwood, ancient cousin of the American sequoia and known only as a fossil in the rest of the world.

Under the trees there's a tangled undergrowth of tropical vine and fern, rich green moss, and bamboo thicket. The Chinese call the species of bamboo that grows here "arrow bamboo" because the nomads of long ago made their arrows out of it. It is also known as Chinacane. It has slender stems, the largest of which are about an inch and a half in diameter. And it thrives in the moist air and heavy rainfall.

Where the towering trees grow thickly, the bamboos are stunted because they do not get much sunlight. But in large sections of the wilderness, the tree cover is sparser. This started to happen long ago when primeval trees were cut down by nomads who wanted to raise maize to augment their hunting. As the centuries went by, the nomads were joined by lumbermen who also felled the forest cover. Free of shade, the bamboos flourished, forming a wall some ten to twelve feet high, with stems so thickly crowded together that it is almost impossible for human beings to penetrate. Through this jungle meander trails of all sizes, worn there by the creatures who inhabit the area.

Above 12,000 feet, the bamboo thickets and the forest trees die out and rhododendron bushes take over. Higher still, the belt of rhododendron also disappears. At this altitude there are only open alpine meadows broken by barren cliffs and steep rock-strewn slides, or screes, that stretch upward to the snow peaks.

This wilderness with its various belts is the home of many strange and rare creatures. So precious are they to the People's Republic of China that in 1965 it created several wildlife preserves in the area. No human being is allowed to enter the preserves without permission, even to gather medicinal herbs.

One of the sanctuaries is Wanglang Preserve, which lies in the mountainous region of north-central Szechuan Province. The closest village to the preserve is Kao Keng, some 8,000 feet above sea level and less than two miles south of the preserve's border.

In Kao Keng, the dozen or so houses that line both banks of a small stream are built of wood. The families who live in them are Tibetans. To keep warm, even the children wear long woolen robes, or *shubas*, that reach to the calves of their legs. Near the village lie fields of oats, buckwheat, highland barley, and maize, for it is much too cold here to grow rice or even wheat. Beyond the fields are pasture lands where the villagers keep a few cows and sheep. They also have beehives. They augment their income by collecting the medicinal herbs that grow in moist places on the craggy mountain slopes that surround them.

The people of Kao Keng feel a special responsibility toward the wildlife that lives so near them. There are species of pheasants here that cannot be seen wild in any other place in the world. White Tibetan snowcocks live high up on the slopes, near the line of perpetual snow. In the upper brushlands are equally rare Lady Amherst pheasants with great green ruffs and graceful blue-brown plumage. Down in the lower forests gold-and-scarlet golden pheasants hide themselves.

Many of the native animals are equally rare. On the alpine meadows Tibetan blue sheep graze. Takin, serow, and goral, all members of the goat family, frequent rocky slopes, alpine meadows, and forests. The leopard and lynx, the Himalayan black bear, and the Tibetan blue bear roam at will throughout this wilderness. Various species of deer graze in open woodland glades and alpine meadows. The lesser panda with its fiery red coat and the black-and-white giant panda make their home in the bamboo jungles.

The giant panda, rarest of these animals, is the most important to the villagers. They are better acquainted with their strange neighbor than any other persons in the world. They call it the "bear-cat."

5: *Prologue*

# 1
# Finding Mates

FOR MOST of the year the giant pandas are silent, but twice a year, in the spring and again in the fall, the deep fluting roar of the males can be heard drifting through the bamboo jungles.

It is April in Wanglang Preserve. High in a tall spruce tree one of these males, a nine-year-old, is perched. He is about six feet long from the tip of his nose to the end of his stumpy tail. And he stands about four feet high when he is on all fours. But now he is just a great unwieldy ball of white-and-black fur crouched in a fork of the tree.

His legs are black, and a broad black band crosses his shoulders. His ears are black, too, and large black circles around the eyes give his face a sad clownlike look. His pelt is a woolly coat of stiff springy hair that lies close to his skin in layers, two inches deep in some places. Beneath this outer coat he has another very thick, somewhat oily undercoat. It is enough to keep him warm in the 8,000- to 12,000-foot altitudes where he spends most of his time. Here the temperature seldom goes above 15 degrees in the winter and averages 65 to 70 degrees in August, the hottest summer month. Usually it is damp and chill the year round.

Though this panda ordinarily weighs some three hundred

pounds, he has lost his customary sleek look, for he has eaten next to nothing in the past few days. All his being is concentrated on finding a mate.

For the better part of the year he lives a solitary life, as all pandas do. Only at the mating season do males and females come together. Then giant pandas may be seen in twos and even threes as they search and spar for mates.

There is a tangled underbrush of bamboo around the tree in which the panda is perched, and some scattered open spaces out of which other trees rise. Occasionally the panda stops his roar to scrutinize the wilderness below him. Though he is rather near-sighted, his sense of smell and hearing more than make up for his poor vision.

He knows that if he waits long enough and calls long enough, somewhere in that jungle a female panda will hear him and come his way. The wind will bring him her scent and the sound of her movements through the bamboo, and he will clamber down to join her.

At the very moment the giant panda is roaring in his treetop, a female panda is also starting on her quest. She lumbers along one of the trails. At two hundred pounds she weighs somewhat less than the male, and her ponderous body resembles a great stuffed toy. She walks flat-footed as does a bear. But her stride is longer. Her gait is a kind of pigeon-toed shuffle, and she sways from side to side, her head bent low, her stubby white tail pressed against her body. For several days now, the female has been eating even less than the male. But despite her fasting she is quite energetic, pressing forward doggedly while emitting sheeplike bleats, *he he he.*

Though she is large, she moves like a shadow through bamboo thickets that are difficult for human beings to penetrate. The way

7: *Finding Mates*

leads upward steeply, skirting narrow ledges, sometimes taking a sudden plunge downward into dark ravines. But the female panda travels with surprising agility.

When she comes to the banks of a chattering, ice-cold stream, she stops to lap at the water, then fords it. She will cross several more small rivers before she reaches her destination.

She isn't carrying out a haphazard search for a mate. This route is familiar to her. For four years, ever since she reached the mating age at six, she has followed it. She knows instinctively that somewhere along it, a male panda will be waiting for her.

This year, as in every other year, she leaves behind signs for any male panda to read. She does this by raising her shoulders, bending her head even lower, and lifting her stubby tail to press her rump against a rocky ledge or a tree trunk. She rubs back and forth, depositing a small patch of glandular moisture. Though the substance is almost odorless to human beings, it gives off a scent that will attract the male to her.

Every now and then the panda stops to listen, her head cocked to one side, her short ears pricked up. From far away comes a low, deep call. When she hears it she always stops to leave another scent, then continues on her way.

Each time the call seems closer. At last it is quite loud. She is approaching the old spruce tree where the male panda is sitting. Her odor drifts up to him, and he grows excited. Quickly he slithers down the tree, rump first. When he is within a few feet of the ground he jumps down and hurries off in the direction of the scent and the bleating.

He is eager to mate, and as he nears the female, he gives a loud roar and starts running toward her. But the female's mating time has not yet come, and no mating can take place until both animals are equally ready.

8: *Finding Mates*

As the male panda comes closer, the female turns and runs from him. The male chases her and presently overtakes her. Immediately he tries to mate with her, but she backs away in fury. Her great paw strikes out at him. Her long, sharp claws rake across his shoulder drawing blood, leaving behind a jagged flesh wound.

Claw gashes such as this are common among adult male pandas, and many bear the scars of such encounters. But a little thing like a clawing will not stop the male panda. During the days that follow he continues to trail the female, making yodeling calls to her, which she ignores. When he gets too close, she turns and lashes out. Again and again she draws blood, but he will not give up until his mating period has passed. It will be a very short period, usually no more than ten days. If the female panda is not ready before then, the male will be unable to mate this spring. He will have to wait until his capability returns during the fall mating season.

This male panda is not the only one to pick up the scent of the female. Suddenly one morning another roar comes from the bamboo thickets. A second male eager for mating is on the trail.

It is too much for the female. She runs for a tall spruce. Hugging the trunk by pressing the soles of her feet against it, she shins up, humping her back like a caterpillar while her claws dig into the bark leaving long gashes. Now and then she snatches at branches and snags from broken limbs to support herself.

At last she reaches a comfortable crotch between the trunk and a high branch. She settles into it, curling up in a snug ball. There, chattering her teeth in excitement, she watches the male panda below her. He is busily stalking about the area, marking rocks and tree trunks with his own scent. Sometimes he stands on his front legs and backs his rump high up the trunk of some tree to leave a scent there too. He wants it understood by every visitor that this is

*9:  Finding Mates*

his territory and the female is to be his mate. If any other male panda enters it, he can expect a fight.

He hasn't finished his marking when a second roar sounds close at hand. Out of the bamboo thickets plods another panda. He sniffs at the scent of the first male, but it doesn't frighten him off. The attraction he feels for the female perched in the spruce tree is too strong for that. He is going to make a challenge.

On he comes. The first male goes to meet him. Swinging their huge paws, they swat at each other with resounding blows. They snap and bite with their great teeth. Back and forth, round and round they go. Finally the second panda begins to back away; soon he slinks off into the bamboo thickets. The victor celebrates his triumph by rolling around on the ground.

Early the next morning the female panda, having reached her mating peak, slithers down the tree. Now it is her turn to do the chasing. But she does it backward, presenting her rump to the male panda. As she comes toward him, he backs away to the tree, where he sits down propped against it. When the female reaches him he clasps her around the waist and holds her tightly, giving a loud cry. And so they mate.

The actual mating lasts only a couple of minutes. When it is over the female panda turns on the male, tired of his company. She gives him a hard bite that warns him to be on his way. He scrambles to his feet and ambles off. The tie of nature that has drawn them so closely together for a brief spell is broken.

No sooner has he disappeared than the female forgets about him. She is very tired; the activities of the mating season along with her long fast have worn her out. Sluggishly she makes herself a nest in the bamboo underbrush at the foot of the tree, turning round and round until she has formed a circular bed of crushed bamboo stalks. On these she stretches out, flat on her back. In no time she is fast asleep.

10: *Finding Mates*

# 2
# Wanderer

LATE in the afternoon the female giant panda wakens in the cool green shadows under the tree. She stretches once, then slowly staggers to her feet. From somewhere in the distance the fluting roar of a male panda comes to her ears. But she no longer responds. She feels only a faint stir of hunger. Wandering into the bamboo thickets, she searches out some tender bamboo shoots and begins to nibble on them. At present only a few bites will satisfy her. But in the weeks to come, her appetite will quickly return. A month from now she will be eating more than ever, for she is carrying new life within her.

Like the bear, the giant panda is classified as a carnivore, or flesh eater, because it will occasionally devour small animals or birds. But its main diet is made up of plant life, principally the leaves, shoots, and stems of the arrow bamboo. Except for its tender shoots, bamboo is so tough that only an animal especially equipped to handle it could survive on this diet. The panda is such an animal. Over centuries it has developed a massive head with strong jaw muscles and huge teeth, the upper surfaces of which are deeply ridged.

With these great teeth the panda bites off a bamboo stem close to the ground. She picks it up with her handlike front paw, which appears to have six digits rather than five. Actually the sixth claw is only an elongated part of her wristbone covered by a tough pad of flesh, which acts like a thumb. It is muscular and supple so that she can easily handle even the slenderest stalk.

She lifts the stalk to her mouth and strips off the tough outer sheath with a sideways jerk of her head and a twist of her paws in the opposite direction. Then she places the stalk lengthwise between her cheek teeth and bites off a chunk. Slowly and methodically she chews on it with those great grinding molars.

But even massive molars cannot do a thorough job on the tough bamboo, and the panda swallows bits and fragments whole. The sharp bamboo slivers proceed down her esophagus without harming it because it has a horny lining. Her stomach walls, too, are thick and muscular, resembling a chicken gizzard. But the intestines, which should have been long enough to digest the bamboo, are instead very short, only about five and a half times the length of the panda's body. This is short even for a carnivore, and much too short for an animal whose main diet is green stuff. No one can explain why the panda's intestines did not develop as did its jaws and teeth, esophagus and stomach. But because they are so short, much of the nutrition in the stalks passes out of the panda's body. One of the most obvious signs of a panda's presence are numerous large droppings that contain quantities of undigested bamboo.

To sustain herself, the giant panda has to spend as much as twelve hours a day eating. Her intake may be as high as twenty pounds of bamboo daily. Sometimes she stays in one spot long enough to gnaw off all the bamboos in an area some sixteen feet square; at other times she eats on the move.

After feeding, the panda places her front paws before her face and licks them clean, going over her forearms too, and finally wiping one paw over her face as a cat does. If a bamboo sliver sticks between her teeth, she uses the claws of one of her front paws to pick it out. Occasionally, perhaps to sharpen her claws, she strips away the bark from the entire lower trunk of a tree, up to a height of several feet.

The panda has no set schedule, though she sleeps more at noonday and midnight than at other times. She rests when the mood is on her and in almost any position you could name— sitting up with her back against a tree, lying flat on her back with her legs in the air, or curled up in a ball. She spends a great deal of time, too, just twisting and rolling on the ground for exercise, scratching herself and licking her fur. For all her size she can twist herself into so many strange postures that she looks like a rubber animal.

Sometimes she stands up on her hind legs like a bear, especially when she wants to reach for something. All pandas do this. But nobody has ever seen one take a single step in that position. Perhaps this is because the panda's bones are heavier than those of the bear, which sometimes does walk upright. The added weight may be too much to support on the panda's hind legs alone.

One of the panda's favorite positions is to sit with her back against a tree, one hind leg lifted straight up in the air. With her chin propped on her raised foot she sits staring off into space. Occasionally when she's dreaming away like this she is interrupted by a swarm of tiny brown birds whose territory she has ignorantly invaded. Scolding and chattering and fluttering around her angrily, they finally drive her away to find a quieter retreat.

Everywhere the birds are sensitive to intrusion this time of year, for it is spring and their mating season too. Their cries and

13: *Wanderer*

songs assault the panda's ears. All night long the *huuuh uh huuh,* of the great tawny owl can be heard as he howls out his welcome to the season. But when he is courting he adds an embellishment, a quick trill with an *oo. Kyuwitt, kyuwitt,* his mate replies.

Other birds are also giving voice to the usually quiet wilderness. Yellow-eyed flycatchers and gregarious bubuls in scarlet topknots call to their mates at intervals. The little Szechuan cuckoo's strange cry floats through the woodlands day and night. Most melodious of all is the song of the black, red, and gold Peking robin, the Chinese nightingale with his clear sweet melody; his *tee tee tee tee* floats out from the spruce and pine forests which are his home. Sometimes the panda can even hear the loud melodious whistles of the mating Tibetan snowcocks drifting down the mountainside from as far as a mile away.

Once in a while a shy russet-colored civet darts in front of the panda in pursuit of a centipede. The civet, cousin to the mongoose, looks something like a cat but with protruding eyes, shorter legs, and a longer muzzle.

Occasionally the panda's path crosses those of the Himalayan black bears, "dog bears," as the villagers call them, with their pointed snoots and the broad white crescents on their chests. The black bears, usually solitary like the panda, are now seeking mates, as are the rarer Tibetan blue bears, which the villagers call "horse bears." They are larger than the black bear or the panda and have pale straw-colored coats. They are very aggressive, and their mating period is a fearsome thing.

Seen together, bears and giant pandas look as if they might be related species, though the giant panda has a much larger and rounder head than the bear. The skeletons of both animals, the arrangement of their muscles and the shape of their brains, are

*14: Wanderer*

similar. Both species have short and stubby tails and both shimmy up trees and descend them rump first.

Now and then at dusk the giant panda meets another possible relative, the lesser panda, wandering through the wilderness. Unlike the giant panda, the lesser panda is a sociable fellow. He usually chooses one mate and lives with her for life. Together they raise a litter of cubs every year in the hollow of some old tree.

The lesser panda is only about two feet long, with a bushy ringed tail that adds another eighteen inches to its length. It is clothed in shining chestnut-red woolly fur that gleams in the twilight. The black markings about its face and ears, the white spots over each eye and on the tips of its ears, give it a pixieish expression.

Like the giant panda, the lesser panda feeds mostly on bamboo. It, too, is equipped with powerful jaws, huge teeth, and handlike front paws, though it doesn't have a sixth claw. It also has very short intestines. These are some of the reasons many zoologists place the giant panda with the lesser panda in the raccoon family. Others maintain that the giant panda and the lesser panda are not related but only developed in a similar fashion because their ancestors both took to eating bamboo.

Besides bamboo, the lesser panda eats fruit buds, leaves, insects, and larvae and is not above robbing birds' nests of eggs. It doesn't much care for meat, but it will occasionally grab up a mouse as it skitters across a clearing.

Usually the lesser panda spends the day in sleep, curled up like a dog or cat on the branch of a tree with its tail over its head, or with its head tucked under its chest and between its forelegs. When a male lesser panda talks with his mate it is in a series of short whistles or squeaking notes. Occasionally he emits a loud *wha wha wha,* sounding so much like a small child crying for its

15: *Wanderer*

mother that among the villagers he's earned the name, "child of the mountains."

Traveling through the open clearings, the giant panda often catches the sounds of timid creatures warning their neighbors of the presence of nearby danger from their natural enemies, such as the wild dog or the leopard. At the warning sneeze of the leader, a pack of tawny musk deer skitter off for safety with rumps and heads strained high, half galloping, half jumping along with all four feet off the ground.

The doglike yaps of the shy muntjac, or barking deer, signal its fellows to disappear into the bamboo thickets, where it is almost impossible for a leopard to spring. The panda may also hear the warning cry of the dark woolly sambar, an antlered deer, whose loud whistle sends the herd off with flashing heels and a flicking of tails revealing white undersides.

When the panda senses danger herself, she has two choices. She can melt into the thickets or shimmy up a tall tree where she crouches on a high fork. Here, hiding among the thick leaves, she clicks her teeth together in a chittering warning that sounds like twigs being rubbed together in a wind.

Treetops are one of the panda's favorite resting places anyway. Here she is safe from prowling leopards and packs of wild dogs that are as rapacious as wolves. From her high perch she can look down dreamily on a silent world. Long, pale green-gray streamers of moss festoon the tall spruce trees and sway in the wind. Beyond the clearing, bamboo thickets stretch away on all sides, rising in green phalanxes toward the distant snowy peaks.

In the open space below the panda's perch, a rust-colored lynx, a little larger than an oversize cat, steals across the bed of soft moss and springs on two green monal pheasants engaged in their courting dance. Just in time the birds scatter into the brush,

*17: Wanderer*

piping loudly. Of as much danger to them is the great eagle soaring overhead. The eagle is a carnivore with hungry eaglets to feed and no bird, vole, pika, or meadow mouse is safe from it; nor are fawns or the kids of the mountain goats, blue sheep lambs, or

*18 : Wanderer*

the piglets of the wild boar. With its sharp talons and tearing beak, the eagle can overpower creatures far larger than itself.

Certainly the eagle would have no problem with a newborn cub of the giant panda. But the adult panda need have no fear of eagles. She is too big and strong to be troubled with many enemies. Bears offer her no real threat. Though she may spat briefly with them over territory, she usually avoids them. Shaggy wild dogs that roam the wilderness and look much like German shepherds with their plumed tails and lean faces are much smaller than she. But a pack of them could be dangerous if they managed to corner her.

Another threat is the snow leopard. This handsome cat, found only in the Himalayan, Tibetan, and Altai mountain ranges of Asia, is easily the most beautiful cat in the world. Some four feet long, it has a tail almost the same length, and for most of the year it wears a creamy coat of thick fur decorated with dark rosettes. It roams the mountain slopes above 8,000 feet, seldom dropping below that altitude. It follows the life of a hunter, preying on any creature that it can surprise and kill, from wild sheep to goats and deer. Snow leopard young are born in the summer; to keep them fed the adult leopard becomes very bold indeed.

Face to face with the giant panda, the snow leopard would be no match. A loud bark from the panda, a swipe with one massive paw, and a ferocious snap of the great jaws would easily drive off the leopard. But the big cat hunts by stealth and depends on a sudden spring and a well-aimed strike at a vulnerable spot. The panda cannot rely on her poor eyesight to warn her of danger, but has to depend on her acute senses of smell and hearing to alert her to this sly enemy which, when famished, will take any chance.


*19:  Wanderer*

# 3
# Alpine Meadows

THROUGH the quiet wilderness the giant panda moves like a great black-and-white shadow. The dense bamboo thickets mounting up almost perpendicular mountain slopes rustle with every breath of wind. As she plods along, dust from dead leaves blows into her eyes; every now and again she must stop to rub at them with a paw.

Though she seems to be wandering aimlessly about, she is actually following a meandering course that will eventually lead upward. As summer comes, the lower forests are too hot for her. At 70 degrees, especially if it is humid too, she becomes listless. If the temperature rises higher than this, she pants and her pulse races. She loses energy and appetite. If she were to stay in the heat, she would sicken and die.

Summer brings the rainy season, and winter's snow disappears into muddy slush. Frequent downpours send torrents streaming through the jungles. The whole world seems to split apart as huge avalanches of mud, boulders, and tremendous rocks thunder down to the valleys, destroying everything in their path. But the giant panda is used to these violent outbursts and continues on her way undisturbed.

Sometimes, instead of rain, chill mists and fogs drench the bamboos, showering the panda with drops of moisture as she passes through. Underfoot, thick green moss forms a deceptive coverlet over everything, concealing decaying tree trunks, beds of rotten bamboo stalks, sharply tilted slabs of rock, and other pitfalls. Moisture and occasional ice make the moss slippery. But the panda moves easily over the glassy surfaces, even up the steepest slopes. Hair growing on the soles of her feet gives her traction, and her claws help her to keep her footing.

As she mounts the higher slopes the wind becomes very chill, sometimes bringing snowfalls with it. The panda rejoices in the cold white powder under the trees. She rolls about in it and presses handfuls of it against her face before continuing onward.

All around her, the small creatures that live in rock clefts or underground burrows or hollow trees are still conducting their mating rituals. The young of the shrew are easy game for the panda, as are meadow mice and the long-tailed harvest mice and slate gray field rats. So is the pika, which looks like a tiny rabbit with short, rounded ears. As the pikas work gathering greens for the winter, their short bleats sound over the mountain slopes, signaling to one another, perhaps warning of danger. At the first hint of danger they'll pop out of sight into burrows or rock crevices. But occasionally one of them is careless—a sad mistake when the giant panda is around. Out darts a huge paw, breaking a bleat midway through, and a tiny pika disappears into the great jaws. It will have no use for its summer harvest now.

As the panda pushes forward through a woodland glade, life explodes above her with a burst of chattering voices, welcoming, mocking, scolding. Large brown eyes follow her movements with a glittering stare from their high tree perch. These are the rare golden monkeys, whose beautiful thick fur ranges from snow white in the babies to a pale gold in the females and rich russet bronze tones in the males. Bright blue, turquoise, or violet patterns spread across their upturned noses, like butterflies perched there with outstretched wings.

In their thick coats, the golden monkeys, like the pandas, are unable to stand the summer heat. Now this company of eighty or so is headed for the upper regions and has stopped along the way to feed. They nibble on tender leaves, ripening berries, insects, and eggs, and their trail is strewn with empty acorn shells. Their appetite satisfied for the moment, the monkeys continue on their journey, flying and swinging through the treetops. The giant panda is alone again.

Her trail leads along the margin of a steep slope that falls away to a rushing torrent some five hundred feet below. Suddenly she hears the *clomp clomp clomp* of hooves behind her. The mighty takin that has been wintering in the valleys is migrating upward too.

The takin is a strange animal found nowhere else in the world. Some scientists place it with the musk ox, which is a cross between an antelope and a goat. Measuring from the shoulder, the takin is more than four feet high and weighs up to seven hundred pounds. Its short sturdy legs and feet are goatlike, but with huge hooves often six inches in diameter. It has a long face with a humped nose and a broad muzzle, and its head is crowned with curved horns. Its coat is dense and shaggy.

During the spring the takins have been grazing in the valleys. But with the approaching summer they begin gathering at the natural salt licks on the lower mountain slopes. Over the years the takins have beaten well-worn trails to these licks. By June, large herds are milling around them.

The herds split up into companies of a hundred or so, each with its own old bull leader. The companies set off along the wide trails that lead to the alpine meadows. Despite their size—they are almost two and a half times heavier than the panda—and clumsy appearance, takins are surprisingly agile. When frightened or enraged they will thunder down on the offender with flashing hooves and lowered heads, so swiftly that it's difficult to get out of their way.

Now, one after the other, the great creatures come up the trail toward the panda. She knows better than to confront them and chooses the simplest course out of her predicament. Up go her front paws to cover her eyes, and over she rolls into a great furry ball. The "ball" gives a bounce, and over the edge of the steep slope tumbles the panda. Down she goes without taking so much as a peek around her. When she reaches the margin of the stream, ice-cold spray showers her. Only then does she uncoil, shaking herself.

A large gray water shrew mole, startled by the racket, pokes his head out of his burrow to take a look. But the panda ignores him. She is very thirsty, and the water is refreshingly sweet. After drinking her fill she wades into the torrent and easily swims across it, stopping at the far shallows to scoop up a small fish and pop it into her mouth. Then she is on her way again, plodding up the steep bamboo-clothed opposite slope.

Day after day she travels in this leisurely fashion. Each slope she climbs brings her higher into the mountains. Almost every jagged ravine she crosses is filled with a torrent of whitish green water. Presently she emerges from the dense bamboo jungle into

23:   *Alpine Meadows*

more open thickets under a tree cover of towering spruce and lichen-patterned alders.

Here the panda sometimes meets up with the oddly fashioned serow, a member of the goat family found only in Asia. Brownish black, with horns close to ten inches long, the serow sports a beard from the corners of its mouth to the base of its huge ears, giving it a whimsical, melancholy expression.

The serow prefers to live in thick spruce forests, and it stays in practically the same territory year round. On the other hand, its smaller cousin, the goral, occasionally crosses the panda's route while roaming across the rocky screes. The goral, which also lives only in Asia, is a graceful olive-buff animal about the size of a goat. Its most amazing feature are its hooves, which have shallow depressions that act like suction cups. The main digits and dewclaws also end in black rounded rubberlike soles, which enable it to keep a firm footing on sloping rocks. With these the goral can bounce easily along on ledges that are so narrow even a rabbit would have difficulty perching on them.

Beyond the bamboo thickets the panda enters the belt of low rhododendron trees and bushes that in late May and early June blaze with multicolored blossoms—reds and pinks, whites and yellows. Rhododendrons are evergreens, but because the ice and snow linger so long in these regions, their leaves are warped and curl up in tubelike shapes.

Still higher she climbs until the belt of rhododendron also dwindles away. Now there are only open alpine meadows broken by barren cliffs; steep rock-strewn slides stretch upward to the perpetual snowpeaks, gleaming like crystal against the clear blue sky. Summer warmth strews the rich meadow grass with large red, blue, and golden mountain poppies, primulas, and tiny blue

*25 : Alpine Meadows*

lilies. But even summer weather is uncertain. In late July and early August, the hottest season, the temperature averages only 50 degrees. A sudden blizzard may sweep down over the meadows and bury the green grass and flowers. When that happens, only a few tiny lilies or golden poppies encased in ice remain to remind the frozen world that spring has really come.

This is the home of the wild blue sheep, the bharal, that live at 12,000- to 17,000-foot elevations. They are neither sheep nor goats but spring from a common ancestor that existed in the Early Pliocene era in China. Their faces are long, their eyes large. Small pointed ears and curving horns sprout from their heads. Clothed in grayish blue hair, they can stand motionless among the rocks, hidden from enemies by their natural coloration.

The blue sheep share the meadowlands with the rare Thorald's deer who carry great branching antlers on their heads. Both blue sheep and deer have sentries to warn them of danger. But they ignore the panda venturing into their meadows, as she ignores them. She is attracted to the wealth of succulent plants that the summer sun has brought up. Along the banks of the streams that course through the meadowlands grow gentians, irises, crocuses, Chinese vines, and tufted grasses. The panda grubs them up and devours them in quantity—they are a change from her common diet of bamboo and occasional rodents. But she is uneasy in these open lands where the cunning leopard roams at will, so she never strays far from the tree cover.

By late August, the period that the Chinese call the Time of Cessation of Heat, night frosts begin to shrivel plants and grasses. Snowstorms become more frequent and fiercer, ruffling the thick fleece of the blue sheep.

The giant panda turns back into familiar territory, the sanctu-

26: *Alpine Meadows*

ary of the bamboo thickets. Her swelling abdomen is plainly visible now. She is normally so round and large that if she were going to have just one cub it wouldn't be obvious, but the size of her abdomen indicates she will be bearing twins or even triplets, though triplets are rare among pandas.

Her breasts—she has four of them, one pair on her chest, another on her abdomen—are swelling too. Her appetite is gradually falling off; all she craves are bamboo stalks. But as the days go by she eats fewer and fewer of these. She knows from the growing weight within her that her cubs are about to be born. She must select a good den in which to give birth.

The black bears and the Tibetan blue bears also are looking for dens in which they can hibernate for the winter. They are sleek from their summer of stuffing in preparation for their long winter's sleep. The pregnant females will waken to give birth to their young during this hibernation period; then, with their babies safely suckling, they will return to their half-sleep again. When they emerge the following spring they will be accompanied by their cubs.

The giant panda does not hibernate, but the female must hole up for a while during the birth and the early weeks of her cubs' existence. For this she needs a place that is both comfortable and snug.

At last, on a steep slope covered thickly with bamboo, she comes upon a kind of cave formed by an overhanging rock. Old droppings around the cave show that it has been used often before. It is this panda's special birthing place to which she has returned year after year. The cave is satisfactory in every way. It is roomy and well concealed. The bamboo thickets that grow around it will provide her with food, and a small stream sparkling in a nearby gully will satisfy her thirst.

27: *Alpine Meadows*

But neither food nor drink interest the panda at present. Now she has stopped eating altogether. She seems to have lost most of her energy too, as she languidly moves about preparing things. She cuts down bamboo stalks and drags them into the cave to form a nest. When she has finished she spends most of her time sitting in her den or sprawled outside resting. For long periods she lies in a heavy sleep—waiting, waiting.

28 : *Alpine Meadows*

# 4
# The Cubs Are Born

THE DAY of birth finally arrives. The panda knows her time has come and crouches in her nest, feeling her belly moving up and down in heavy rhythmic movements. It contracts and relaxes, contracts and relaxes, as though it were being squeezed at intervals by a gigantic invisible hand. At first the contractions are few and far apart, but as time goes by they increase in number and intensity. The giant panda helps the process by pushing down with her abdominal muscles.

Suddenly a cub is born. Head first, it drops to the ground, a squirming grublike creature weighing only four or five ounces, no bigger than a small rat. It is difficult to believe that it will ever reach the size of its two-hundred-pound mother.

Its eyes are tightly shut, and it is toothless. Its body, which has a sparse sprinkling of short white baby hair, is a light pink color. Its tail is about one-third its body length, and its head is rounder and blunter than its mother's. It is a male, though it is impossible now to judge its sex. Even after it has matured at six years of age it will be difficult to see its genitalia because, as in all pandas, male or female, they will be concealed in protective layers of fat.

Only at the mating season do the male sex organs protrude from that fleshy pocket and become plainly visible.

Now the tiny creature squirms on the ground helplessly. Twenty minutes after its arrival a second cub, a female, is born. Her sex is no more apparent than that of her brother.

The exertion has worn out the mother. She is panting hard, her breath coming in short, uneven gasps. Even though there is nothing in her stomach she begins heaving with nausea. She is very weak, but she props herself up in a sitting position and tenderly picks up the tiny cubs one at a time with her mouth and enfolds them in her arms.

The little creatures are covered with a sticky birth substance, and she starts licking them clean. She licks for two hours before she is satisfied. Then she raises them to her nipples, and they settle down to nurse. The mother panda falls asleep with the cubs suckling at her breasts.

Outside the dark cave world of the panda, autumn is taking possession of the mountains. Frost touches birches, poplars, and chestnut trees, and they burst into colors of bronze and gold and flame. Those creatures of the wild who are stirred by fall mating instincts begin their courtship games.

Male musk deer, usually timid, meet each other in fierce challenge. Entwining their necks around each other to get a good hold, they gash into hide and muscle with their rapier-sharp tusks. The fights are so fierce that before the season is over all the males will bear new scars to join the crisscross of old ones.

The red deer that live in the wooded valley bottoms, the gorals, and the blue sheep are also dueling for their right to mate. Wild male boars, who have been living alone, grubbing and eating their way through the forest, forget about food and the pleasures of a solitary life and begin trekking for the mating places. Here the huge boars fight one another with their venomous tusks until they have won their females. Then round and round in a circle the male drives the female, boxing her roughly with his snout and emitting loud rhythmic grunts.

Out in the forests the lesser panda is also courting, but in a gentler style. Its *wha wha wha* floats through the forest and is joined by the mating call of the giant panda. Sunk in a deep sleep, the female panda does not even hear it.

For a day after the birth of the twins, the giant panda continues to rest and sleep in her den, hugging her cubs to her. When they are hungry they cry out like babies or yap like young puppies, their voices astonishingly loud for such tiny creatures. Then their mother wakens and with paws and mouth gently guides the little mouths to her nipples. When they settle down to nurse she falls asleep again, only to be awakened an hour or two later. In the weeks ahead, the cubs will nurse six to twelve times every twenty-four hours.

The constant feeding has its effect; as the days go by, the cubs start filling out and losing their ratlike appearance. About a week after their birth the hair on their ears and around their eyes and shoulders starts turning gray. The grayish color soon spreads to their front limbs and darkens on their ears and around their eyes. Within two weeks their hind limbs as well as their front ones have turned black, and the black rings around their eyes have grown larger. Though their eyes are still shut, the rings make it look as if the tiny creatures are wearing black spectacles that give them a roguish expression.

By the sixteenth day their chests start turning black, and the black circles around the eyes change into long, slanting splotches like those of their mother. In the next few days the neck, front, and back have turned black. Within three weeks the color has spread over the whole neck and chest, while the white hair has grown longer. Now they look like miniatures of their mother except for their rather long tails and the lighter color of their black markings. They won't turn a deep ebony until the fiftieth day.

During their first month the cubs' eyes have remained closed; they have no reaction at all to light, even when sunbeams slant into the den striking against their closed lids. Only by the end of that first month do they start showing sensitivity by wrinkling up their eyes when the light falls on them.

One day, the male cub manages to open an eye halfway for an instant. The light is too much and down blinks the lid. Immediately he opens the other eye, also halfway and just for an instant. It will be two weeks before he will be able to keep both eyes open at once.

Ten days later, his sister's eyes are fully open too. But even though both cubs' eyes are now open, their vision is still so poor that everything is a blur. The blur will clear away gradually in the weeks ahead, and by the time they are three months old they will be able to see more plainly.

In the beginning, however, they don't have much use for eyes. They spend most of their time sleeping and nursing. Their muscles are so weak that, try as they might, they can't pull themselves up to their mother's nipples but sway from side to side, whimpering. Their mother sits down when one or the other of them cries out like this. With her front paws she picks up the hungry cub and places it gently on her stomach. From there it crawls to one of her nipples and begins suckling, sprawled out like a human baby.

As the weeks go by the male cub begins to outstrip his sister in size and strength. Even at birth he was bigger and stronger. He is also more assertive. He cries the loudest, catching their mother's attention first. Only after she has helped him to his

33: *The Cubs Are Born*

feeding does she realize the second cub is whimpering too. By the time the smaller cub starts nursing, the larger one has already greedily gulped down a good share of his milk. When he finishes all of it, he shoves away his weaker sister to get at her share too.

This is what usually happens when a panda has twins. The stronger of the two attracts more attention, while the weaker one is unintentionally neglected and often dies as a result. Instead of fighting back, the little female cub just whimpers a little and falls asleep only half full. All this shows up in the weight of the pandas at the end of the first month. The male cub weighs four pounds, the female only three.

Life isn't just nursing and sleeping for the panda family in their den. The mother panda likes to have some fun with her cubs. She nuzzles them affectionately and sometimes hugs them tightly to her breast. When they become impatient and begin to wriggle, she calms them down by stroking their furry little bodies with her big paw. As they grow bigger she plays a little game with them. Back and forth, back and forth she gently tosses them, one at a time, from one arm to the other, almost like a human mother rocking her babies.

During the month following her cubs' birth, the giant panda ate very little. But as the young pandas' appetites grow, their mother has to provide them with greater and greater quantities of milk, and she begins to feed voraciously. Soon she is taking more than one and a half times her normal amount of food, still mainly bamboo.

When she goes out to feed she takes her cubs with her, carrying one in her mouth and hobbling along on three legs with the other cub clutched in her fourth paw. She eats sitting or lying down with the cubs resting on her belly.

Beyond the gloomy den lies a strange and delightful world,

dressed in the sparkling icicles of late fall. Glittering ice coats every hair on every leaf in the bamboo thickets. When the mother goes to drink she finds a thin sheath of ice over the little stream. She has to break through it with a massive hind foot to get at the water underneath. When, after drinking her fill, she turns back to the bamboo thickets, the icicles that have formed in the thick fur of her foot chink and chitter until they melt in the sun.

By the time the cubs are two months old they stop suckling at night, though their mother's milk is still their only food. They no longer cry like babies or yap like puppies when they're hungry. The larger they grow, the softer their voices become, so that presently the only sound they make to get attention is a low, bleating *e-e*.

Within two and a half months they are able to put weight on their hind legs and even take a couple of steps. By the third month the largest cub, who now weighs more than eleven pounds, can walk almost four feet at a time before toppling over. His smaller sister, who weighs only nine pounds, can scarcely go three feet without falling.

Now that they are bigger, their mother likes to romp with them occasionally in the soft snow drifts, batting them lightly with her great paws and rolling them over and over like balls. She always ends the game by hugging them to her and letting them nurse.

The cubs like to tumble about with each other, too. But most of all they enjoy playing the toboggan game on their mother's back. While she is eating they haul themselves over her haunches and up onto her shoulders. It is a long distance for their weak-muscled arms. They use their sharp claws to clutch at her fur for support. When their strength gives out they just hang there, catching their breaths and resting against the soft springy pelt. Finally they reach the top; they let go and come rolling down. They clamber up again and roll down again, over and over. It's more than fun. It's good safe exercise for their muscles.

These are dangerous times for all the smaller wilderness creatures because predators, gaunt from the winter's scarcity, are wandering the land. Their footprints are visible everywhere. Those of the lynx are small and set close together. There's a longer stride to those of the snow leopard. And there are the jumbled tracks of the wild dogs which, though now an uncommon sight in the preserve, are still around. Overhead is the eagle, a silent predator whose black silhouette can be seen soaring against the wintry sky, ready to swoop down on any small defenseless creature in the open.

And finally there is winter itself, spreading a soft white deceptive blanket over everything. Snow bridges formed across deep chasms give way without warning when the sun comes out. Sudden avalanches tumble down steep slopes with a roar and a crash. There are deep pitfalls everywhere.

It is no place for the inexperienced cubs to go wandering around alone. Safety for them lies either near their mother or in the den which she can easily guard. But now, as they grow older, they begin to stray from her. The male cub takes the lead, the female bumbles along behind.

The mother panda has to round them up continually. Sometimes in her concern she carries them back to the den and sets them down inside. But it doesn't do much good. The white wilderness is too fascinating a place to ignore, and as soon as her back is turned, out they pop again. Back and forth, back and forth, like disobedient human children. They never give up trying to outlast their mother's patience—and they never succeed.

*36: The Cubs Are Born*

# 5
# Wilderness Tragedy

BY THE TIME the panda cubs are four months old, they can easily mount their mother's back. There, clinging to her fur, they ride safe above the pitfalls of the path below. This is fortunate because with the passage of weeks the swath of bamboo stumps around the den has been steadily growing wider. The giant panda has to travel farther and farther afield for her fare.

With every yard she puts between herself and the den the job of protecting the cubs becomes more difficult for her. She can no longer shoo them into the cave in times of danger. Instead she has to rely on the scattered spruce trees that dot the bamboo thickets. If danger threatens, she will muster the cubs on her back and shinny up a tree.

At least she no longer has to fear the panda's worst predator—man—who at one time almost exterminated her kind. Man was such a recent enemy the panda had no experience in defending herself against him. For many years the tribes among whom the panda lived never bothered the animal unless it entered their village compounds to rob their beehives. Even then they only drove it away.

To the villagers, the panda was a semi-divinity with its black-and-white human-appearing face and its dexterous front paws. There was a magical quality, too, in the way it moved through the forests like a shadow, despite its huge size. And since, unlike the ferocious bears, it behaved peaceably toward humans, they felt affection for it.

All this changed about a hundred years ago when foreign hunters first began coming to the Szechuan highlands to hunt the panda. At the same time guns were introduced to the area, and high sums were offered for the animal. The impoverished tribes-people, tempted by the money, began to overcome their qualms about killing the panda. They hunted it themselves and also helped Westerners track it through the wilderness.

Using the village hunting dogs to pick up the panda's scent, they would trail the panda. The panda reacted as it did when the wild dogs chased it. It simply shimmied up the nearest tree. Of course it could not know that its only chance of escape was not to let the dogs tree it, but to flee into the bamboo thickets that would hinder the pursuers. Instead, even when it was on the ground it saw so little danger in men that it would walk fearlessly toward them, letting them shoot it at point-blank range.

The People's Republic of China no longer allows man to hunt the panda indiscriminately. Once again, the animal's chief ene-mies are the wild dogs. Banded together in small packs under a leader, these fearsome predators close in on their kill, their short, shrill yaps ringing a dirge through the wilderness.

When the dogs are running, moles and shrews and pikas pop back into their burrows or crevices. Deer and chamois and blue sheep stand motionless, invisible against the rocky screes. But if the wind carries their scent to the enemy, betraying their presence, they are off in a flash of galloping heels and tossing heads.

The larger more powerful animals, the takins, bears, and pandas, when cornered fight with formidable horns, or with strong muscular arms and tearing teeth. Even the plucky little goral can put up a courageous defense. Its back against a rocky buttress, it will turn on the vicious pack and with its sharp horns will gore any careless animal coming too close.

As the panda and her cubs have begun to venture farther and farther away from the birthing cave, the mother has been constantly on the lookout for wild dogs, for she knows that her best defense is an early warning, which will give her time to flee to safety. The weather has been growing steadily bleaker; with the bitter cold the danger from hungry predators increases, for most game is driven back into shelter and out of reach.

On one crisp morning, alert as usual, the mother starts toward the thickets, carrying her cubs on her back. Snow flurries drifting down from the leaden sky dust the pelts of all three pandas. The wind ruffling their fur has a sharp edge. The whole gray-and-white world seems hushed.

Suddenly the mother panda's ears prick up. The sound of yapping fills the quiet day. In alarm, she streaks for the nearest spruce tree. With the cubs still clinging to her back, she clasps the trunk with both arms and starts to shinny up the tree.

But she is already too late. Some eight dogs, the largest leading the pack, burst into the clearing. Their jaws are gaping, baring two rows of sharp teeth.

The panda realizes she can never get up the tree in time to escape those rending jaws. So she slides down again, the two cubs on her back squealing with terror. She whirls, setting her rump against the trunk. Moving with lightning speed, she reaches out with one strong paw and grabs the leg of the leader as it springs sideways toward the cubs. At the same time she pulls the struggling animal toward her. She opens her powerful jaws and seizes the dog's head. With a loud snap she brings her jaws together and the great molars shatter the animal's skull. The other dogs continue to foam around their quarry. The panda's huge clawed paws keep striking out at them. Her deadly teeth chomp. Another dog, scarcely more than a pup, staggers off dripping with blood.

Suddenly the smaller of the two cubs loses its grip and slips from its mother's back. Before she can retrieve it, or even realize what is happening, one of the dogs darts up, seizes it, and retreats. The rest quickly surround him. Soon they are all fighting over the small broken body.

While the dog pack is distracted, the mother panda quickly shimmies up the tree with the other cub still clinging to her back. From a high fork she looks down at the dogs. They have made short work of her cub. Then, still starved, they turn on their own dead and injured comrades. Quarreling and snarling among themselves, they begin ripping the gaunt winter-famished bodies apart, baring their teeth over the raw steaming flesh. Soon it is all devoured. But there was so little meat on the bones that the dogs are only partially satiated. Roused by their kill and by the taste of blood, they rush to the spruce and begin circling its trunk, springing up at the treed panda and yapping shrilly.

On the limb the mother panda paces to and fro in agitation. The dogs, jumping high in the air, are snapping just below her feet. If one or another were to leap a little higher it might be able to drag her off her perch. With the cub still clinging to her back, she again begins to climb—up, up, stopping only when the branches become too fragile to bear her weight. *Hu hu hu,* her melancholy cry ripples out over the wilderness as she clings swaying to her lofty perch.

The dogs presently realize their vigil is useless. They break away, disappearing as rapidly as they came. Their yapping fades into the distance.

The panda waits a little longer. Then she slides down the tree, finally dropping to earth. She is panting from exertion. Retrieving the cub from her back, she clasps it tightly to her breast with one paw. On her other three she hobbles across the snow, now littered with black-and-white scraps of panda fur mixed with bits of tawny hair and bones. Everywhere there is a crisscross of sharp footprints, and scattered among them some bloodstains.

For a few seconds the panda stands bewildered above the blood-drenched panda fur. Then she bends her head and sniffs at it. The odor confirms her loss. She sits down and begins to rock back and forth, giving that familiar melancholy wail, *hu hu hu hu.*

Oblivious to the recent tragedy, the cub in her arms whimpers, *eh eh eh,* and struggles to reach for a nipple. Soon it is suckling. But the mother continues to rock back and forth, wailing. It is as though, despite the evidence of her senses, she is calling to her lost cub to return.

As she wails a steady snowfall blurs all under a soft white coverlet that will presently hide every trace of the tragedy. Overhead the wind is whistling loudly now. Twigs rattle. All around, pine trees and bamboo thickets shed their weights of snow with soft plopping sounds.

The cub finishes nursing and lets the nipple fall. With her mouth the mother gently picks up the sleeping baby by the nape of the neck and heads again for the bamboo thickets, the craving in her belly urging her on. It seems as she plods forward that she has already forgotten both the tragedy and her lost cub. But only because her cub is out of sight is it out of mind. If it were miraculously to appear again, even as long as a month later, she would still recognize it. She would welcome it joyfully with open arms, nuzzling it affectionately and clasping it to her breast for nursing. But, for now, she cannot neglect her own needs and those of her remaining cub if they are to stay alive in this harsh world.

# 6
# The Downward Trek

ON THE high slope where the mother panda and her cubs have been living, the snows of early January shawl the bamboo thickets with a heavy crust, making it more and more difficult to get at the plants below. After one night of freezing temperatures, the mother panda finds the stream has become a pale white ribbon of ice. She stamps at it with her heavy hind foot. But this time she cannot break through. The stream is frozen solid. It is time for her and the cub to descend to lower altitudes since she must find free-flowing water to quench her thirst. Ice and snow alone cannot satisfy her.

She leaves behind signs of her long occupancy. A wide area of bamboo stumps surrounds the den. When spring returns, hundreds of new bamboo shoots will sprout up to hide the havoc she has caused, but other marks will remain. The lower trunks of the trees in the vicinity are all scored with her sharp claws, and some have been stripped of their bark. Most of her massive droppings are buried now under successive snowfalls, but when spring melts the snow they, too, will speak of her long sojourn here, as will the tufts of black-and-white fur snagged on the jagged entranceway to the den.

The giant panda and her cub are not the only creatures in this wild that must adjust their lives to survive the cruel winter season. The Tibetan snowcocks that live near the snowline and the ibis-bills seldom allow winter to drive them far from their summer haunts. The Thorald deer merely grow winter coats of hair twice the length of their summer ones. And the hardy muntjacs even choose this season to conduct their mating duels, like the musk deer, using sharp tusks instead of horns.

Pikas, snug in their burrows or in the crevices of rocks, thrive on the leafy harvests they have collected, coming out occasionally for a breath of fresh air. It is a leaner time for the moles and shrews, which do not hibernate but spend most of their time in their burrows where they live summer and winter, emerging only to hunt.

The wild boars also do not go any distance to escape the winter. When drifts pile up, making travel difficult, they wear narrow paths for themselves through the deep snow. They walk along these single file, moving either against or across the wind in their search for food. But most of the day they spend in the shallow beds they have dug for themselves and then lined with plant materials.

The blue sheep have only a short trek to make. When thick snows on the alpine meadows keep them from the grasses and lichen buried beneath, they descend to the edges of the brush-lands. Here they can graze on the sere leaves and twigs of bushes and low trees.

But the white slopes are etched with the tracks of deer and

chamois and takin who travel greater distances. The graceful goral, which has shed its shaggy coat of hair for long fleece, has an especially hard time in winter. This little creature, so agile in the spring and summer, now moves with difficulty, foundering helplessly in new-fallen soft drifts. At such times it has to plod to some windswept rock to wait, marooned as though on an island, until wind and time have packed the snow and given it a hard crust that will bear its weight.

Only a few of the mountain's creatures can make an easy migration downward. The Lady Amherst pheasants and the Chinese hawfinches can take to their wings to reach the lower valleys where they stay during the bleakest months. Almost as free as the birds in their descent are the golden monkeys that swing through the tops of the trees.

The panda and her cub have to make the long trek on plodding feet over a treacherous landscape. The cub is now well-rounded and sleek, weighing more than fifteen pounds. His tail, which once seemed quite long, more and more resembles his mother's stubby one. Surer on his feet, he is even more active than before.

If there was any lesson to be learned from the tragic death of his sister he has not mastered it. More curious than ever about his surroundings, he frolics at the adult panda's side until some object attracts his attention. Then he's off at a tangent exploring it. The constant *hu hu hu* of his anxious mother rings out at intervals unheeded. Usually she has to go and bring him back bodily. Fortunately, the muscles in his legs are still weak, so he tires easily. He spends most of his time atop his mother's back.

The mother panda follows along the bank of the frozen stream, sensing from long experience that she will eventually come to the place where the water runs free again. The leafy green wilderness

*45: The Downward Trek*

through which she passed in the spring has been completely transformed. Snow and ice now seal over the trails through the bamboo thickets, turning them into long, luminous tunnels. One careless move or jostle and the roofs of those tunnels would come crashing down in a weight of ice and snow.

In other places the snow has bent the graceful bamboo fronds to the ground; ice has cemented them there, forming stiff icy hoops that bar her way. Sometimes she has to descend ice-sheathed slopes so steep that only her hairy soles keep her from falling. Sometimes her path leads across drifts so deep that she has to plow through them up to her belly. Where the snow is packed and firm she makes better time, striding along with a lopelike gait that, though swift, seems leisurely.

Blinding fogs and blizzards sweep down on her. Bitter winds howl from the high peaks that hover over the pandas with a rosy tinge in the afterlight of clear sunsets. Under the full moon the peaks gleam like burnished silver.

The mother stops often along the way to eat and rest and nurse the cub whose only food is still her milk. The little panda was more than three months old when he sprouted the first of his baby teeth—a molar. Now he has two other molars and a couple of front teeth. It will be another two to three months before he has all his teeth.

Though he hasn't enough teeth to chew with, the cub is beginning to show an interest in the bamboo fronds that his mother snaps off and discards. Practicing with his sixth claw, he is soon able to pick them up. While she eats he waves them about and occasionally sniffs at them. It is by its sense of smell that the panda distinguishes edible from inedible food, so the young cub in his play is learning an important skill too.

Along the way the sound of yelping comes once again to the

mother's ears, this time faintly from a distance. She seizes the cub and shins up the nearest tree, peering down in agitation, crying *hu hu hu*. The yapping is followed by several shots ringing out in the still air. Then silence returns again. The panda waits a little longer before she descends. She draws the cub to her breast and nurses it.

Unknown to her, some village hunters and their hounds have been tracking down the wild dog packs as they do every year in the winter, to keep their numbers under control. It is one of the few occasions on which men enter the preserve. Though they have come too late to save the little cub, other wilderness creatures will be safer now.

At last the giant panda reaches a place in the stream where the shell of ice is so thin that she can break through it and get to the water below. She bends her head to drink, and the little cub tries to mimic her. But he has no great love for water, and when his face gets drenched he almost loses his balance and tumbles in. He backs away from the bank, shaking his head in disgust.

A few days later the children of Kao Keng, snug in their fur-lined winter *shubas* and caps, spot the mother and cub at the near bend of the stream that flows through their village. The mother's gulping sounds of pleasure reach the children's ears as she drinks the cold water that rushes along free of ice here. The panda drinks and drinks until her belly swells like a balloon. Sated at last, she turns away, bloated with water. Slowly and awkwardly, the cub at her heels, she moves into the bamboo thickets.

"The bear-cats are back. They're wintering with us again," the children cry.

# 7
# Growing Up

THE VILLAGERS of Kao Keng see the two pandas occasionally during the rest of the winter. They wander through the tangle of bamboo thickets, sometimes in and sometimes out of the preserve. The climate is just right for them, though most human beings would find it too cold for comfort. There are frequent snowfalls, but they are not so deep as on the upper slopes. And though the wind has a biting chill and blizzards are common, they are not so fierce.

There are also days of sparkling sunlight when the little rodents of the wild venture from their burrows to get some fresh air, and the mother panda is able to augment her diet with a taste of jumping mouse or gray field rat. Her cub watches the tiny skittering creatures with interest. Sometimes he chases them across a clearing. But he's too awkward and they're too fast; he always ends up with empty paws.

Often during their treks mother and cub stop to rest. Noon usually finds them sprawled out in a cave or the hollow of a tree or atop a dead stump. But the little panda's siestas are never as long as his mother's. After a brisk game of roll and tumble about with him, she is ready for a good doze, but he doesn't seem at all sleepy.

For a while he sits motionless, contenting himself with shaking his head or his body back and forth in a kind of rhythm, to and fro, to and fro. But he soon tires of this exercise and wanders off to find a snow-covered incline, for he has discovered that it makes a much better toboggan slide than his mother's back. Bleating with satisfaction, he climbs to the top and rolls down sideways, climbs back, then rolls down again. Presently he discovers he can go much faster if he puts his paws over his eyes, curls up, and goes bouncing along like a ball. While only for fun now, this trick will be useful to him when, with his mother, he starts on the upward trek again.

Trees fascinate him. He mimics his mother when she scores the lower trunks with her claws. And he spends a lot of time lying on his back staring up at their leafy tops so high against the sky. One day, when he's five and a half months old, he spreads his arms and legs and digging his claws into the bark tries to shinny up. His first try gets him only a couple of inches off the ground. But he keeps practicing over and over.

One day his mother wakes from a nap to find him nowhere in sight. *Hu hu hu,* she calls.

A small bleating, *eh eh eh,* answers from overhead. She looks up, and there is her cub perched in the fork of a branch six feet above her.

After that, it is impossible to keep him grounded. As he becomes more and more experienced, he mounts higher and higher, completely ignoring his mother's plaintive *hu hu*'s. Among the leafy branches he scampers up and down the limbs so far overhead that if he were to tumble, he would probably break his neck.

But he is not as independent as he seems. He is still easily

*50 : Growing Up*

frightened. One day while he is in the trees a young eagle swoops down to see if he is the right size for eating. Though he is now much bigger than the eagle, it gives him such a scare that with a screech of terror he shins down the tree trunk, plops to the ground, and runs to his mother.

She clasps him in her arms, where he suckles for comfort. At more than six months of age he still depends on his mother's milk, though he now has all his baby teeth. But they're not strong enough or big enough to chew the tough bamboo. For that he will have to wait until his permanent teeth begin to arrive in another six months. Nevertheless, he begins to nibble at the leaves on the bamboo fronds with which he plays. And he practices licking himself clean afterward.

By the time spring comes to the valley in early April no one would recognize in the half-grown cub the tiny ratlike creature that was born seven months before. He is now three feet long and weighs about forty pounds. Despite his big-boned, chunky body, he is amazingly supple.

Spring brings a great delicacy for the pandas. Everywhere in the jungle new bamboo shoots are springing out of the wet earth from under the melting snow. They are tenderer than any other part of the bamboo, and even the panda cub can snack on them. Spring also brings green succulents along stream banks, which the cub can handle easily too. Spring makes possible another delicacy. As the sun grows warmer and the weather milder, the wild flowers begin bursting into bloom. Primrose and cowslips and fragrant lilies strew the meadows. Blue gentians cling to the banks of the stream, and purple violets glow in shady nooks. White butterflies are flitting everywhere. And the tribespeople's bees are humming loudly around their hives on the outskirts of the village.

Those hives, rich with stored honey, are a great temptation to the pandas. To them honey is one of the most wonderful delicacies on earth, and they will do almost anything to get at it.

The young panda's first introduction to honey is a mixed experience. One early morning he goes with his mother to the hives. Following her example, he pounces on a hive and scoops out the honeycomb, smearing the sweet, sticky mess over his paws and face as he smacks it down greedily. But he doesn't know enough to fling himself to the ground and roll about, batting off the bees. Several of them settle on his nose and sting him. With a shriek of pain he waves his paws wildly as his mother rushes to his aid.

Later, after the bees have calmed down and the honey is eaten, mother and cub go over their fur carefully, cleaning off every smear of the sweet, sticky stuff. The cub's painful experience with the bees doesn't dim the lure of honey for him; he comes back to the hives again and again. By now he has learned his lesson and is as quick as his mother to roll about and bat furiously at the angry insects, driving them off before they have a chance to sting.

One day, as mother and cub are loitering near the village, they pick up a tantalizing odor—the odor of roast lamb, more tempting even than the honey. It draws them out of the bamboo thickets and down to the village. They find it deserted, for the villagers have all gone off to gather the medicinal herbs that grow on the nearby cliffs.

The two pandas amble along the street toward the house from which the odor is coming. A push at the door, and it swings open for them. There on the kitchen table sits a leg of roast lamb, cooling while the owners are away.

The mother panda snatches it up in her jaws, and with the cub trailing her, leaves the kitchen and the village. The unfortunate householder returns just in time to see her disappear into the bamboo thickets, the roast clamped between her teeth. That leg of lamb, better than the taste of any vole, pika, or meadow mouse, disappears quickly.

But even honey and roast meat, delicious as they are, cannot hold the pandas in the valley when the season moves on and the days become uncomfortably warm for them. The time has come again to follow the tides of spring upward through the bamboo thickets.

*53: Growing Up*

# 8
# The Separation

THE LAST TIME the panda cub journeyed through the bamboo wilderness all was deep in snow. Now a green jungle full of birds' songs and mating sounds surrounds him. New life is everywhere, the result of the previous fall's mating.

Golden monkey mothers clutch their tiny nursing babies to their breasts. Spotted fawns on wobbly legs suckle at their mother's teats. The musk deer fawns are especially frail creatures and cannot endure exposure to rain and wind. Every time bad weather threatens, their mothers must see that they are carefully sheltered.

The pale babies of the lesser pandas, born blind in their nests in hollow trees and rock crevices, will also be carefully watched over. In some three weeks' time their eyes will open. Presently they will be able to follow their parents around the wilderness in single file.

Female boars, too, each in her own secret birthing hollow padded with thick layers of herbage, are giving birth. The tiny newborn pigs, five to six in a litter and covered only with scant white hairs, gather squealing at their mother's row of teats. Before the year is over, all but two in every litter will have died from sudden freezes and drenching storms.

The young panda's curiosity is piqued by all these strange and interesting creatures. The appearance of some of them is frightening enough to make him keep his distance. When he and his mother meet the mighty takins on their trek upward he doesn't hesitate to roll downhill out of their way. He's curious about the furry little bear cubs that emerge from their caves with their parents at this time of year. But the adult black bears and especially the blue bears, gaunt and irritable from their long fasts, look too mean to approach.

It's different with a peculiar ball of tawny fur curled up and basking in the sun. The young panda can't resist going over and nuzzling it. The response is a snarl and a hiss. Out lashes a paw armed with sharp claws that rake across the cub's nose. He shrieks and swats back. But the tawny ball has unfolded itself into a lynx and streaks away unharmed. The young panda has learned one more painful lesson—curb curiosity, keep aloof from strangers.

He also learns to approach any suspicious object or animal from downwind, so that he can pick up the scent while not betraying his own presence. And he learns how to distinguish among the myriad sounds of the wilderness, for the hearing sense, which is acute in the panda, is as important to it as its sense of smell.

One day, for the first time after many tries, the panda cub manages to snare one of the small rodents that dart across his path. After that he is able to catch an occasional mouse or pheasant chick or pika. Gradually he overcomes his aversion to the rushing torrents he has to ford. And one day he lingers long enough in the shallows near the bank to scoop up a fish.

55 : The Separation

Slowly he is beginning to wean himself. But until he can handle the tough bamboo stalks he will continue to nurse. And when he is hungry he always returns to his mother to be fed. Otherwise, he is more independent than ever.

Like any child, he invents all kinds of new games to play. He works small boulders out of cliff sides and rolls them over the ground. If they go down a slope, he chases after them faster and faster. When they come to a stop, he gives them another push.

In a similarly playful move he breaks off chunks of ice from shaded tree branches or rock shelves and swats them with his paw, sending them flying through the air. He chases the ice chunk, picks it up, and bats it again. With each bat it becomes smaller until finally it disappears in a puddle of water. Then he goes off to get himself another chunk.

Besides Roll the Boulder and Bat the Ice, the young panda does a lot of acrobatics. He practices standing on his big flattish head by leaning against a tree trunk and lifting his hind legs straight in the air. Soon he can do the trick without any support.

Standing on his head, he turns a somersault. At first his somersaults are awkward. But he becomes quite expert with practice and presently he can turn them one right after the other without stopping. Then he starts practicing cartwheels.

When summer is well launched on the lower slopes and the phalanxes of young bamboo sprouts have grown into tough bamboo culms, the pandas make their way to the high alpine meadows. This is a strange land to the cub. Here the newly arrived woolly lambs of the blue sheep, one to a ewe, are trailing their mothers. And the little fawns of the Thorald's deer are gamboling about on wobbly legs.

It is a peaceful scene, but the uneasiness that the adult panda

56: *The Separation*

feels troubles the young panda too. He becomes almost obedient as he follows his mother to the wet hollows and stream banks to browze on the young plants and tender roots that grow there.

Only once does he sight danger in the pale lithe form of the snow leopard stalking through the meadows. He catches a whiff of the killer scent and begins to shiver with fright, cringing against his mother for protection. He is as ready as she when the passing summer turns them back into the bamboo jungles.

Now the cub's independence is almost complete. Close to a year old, he weighs about eighty pounds, a fourth of the size he will eventually become. His permanent teeth are growing out. He is ready for his final lesson—learning how to eat the bamboo culms. Without this skill he can never survive in the wilderness.

At first he is very awkward at it. He rises on his hind legs and grabs at a stalk with his front paws. The resilient stalk sways out of his reach the first several times, and he topples over. But finally he manages to grasp the stem with both paws. He throws himself backward, keeping tight hold of the stem. Then he hooks one hind leg over it and twists around until he is on all fours astraddle it.

Slowly he works his way along it until he reaches the top. Then, holding the stem with one paw, he moves it to his mouth with the other and begins munching on the leaves. He contents himself with only the leaves at first. But before long he is learning to bite off and eat eighteen-inch sections of the culm just as his mother does.

This lesson learned, he is no longer dependent on her. He begins to stray farther and farther away. She lets him go without complaint, and though he may be absent for several days at a time, she no longer feels anxious about him. But she always welcomes him when he returns, clasping him to her and letting him nurse just as he did when he was very small. Actually it is

*57: The Separation*

only an affectionate gesture because now she has little nourishment to offer him. Her milk is drying up.

One day the young panda wanders off and does not return. He is on his own now and will spend the next five years of his life completing the education his mother began for him. His growing-up process will be a time of carefree wanderings. His playfulness will continue. Luminous moonlight will bring him out to feed and frolic in the bamboo glades, a black-and-white shadow flitting through an eerie world. Even during the dark nights he may rouse to eat and play for several hours before curling up for another nap.

But he will engage in some serious business, too. He will start staking out his own territory by scent marking. Rubbing his rump against tree trunks and boulders, he will leave behind his distinctive odor that will warn others to keep their distance. If, during the course of his scent marking, his presence is contested by other pandas or bears, he will fight to keep his place or move on. Finally he will have claimed his own particular section of mountain slope. In this section he will spend most of his adult life, overstepping its bounds only at the courting season when he is in search of a mate. Then he will add his fluting roar to the sounds of the wilderness every spring and fall, summoning a female to his side to create new life with her.

By this time he will have lost his playfulness and become a sedate adult, ambling and eating his way through the bamboo thickets. He will, one hopes, not fall prey to any of the predators that might possibly harm him. A greater threat to him would be some accidental fall from a cliff or a tree which, if serious enough, could cause brain damage resulting in convulsions and an early death. Severe infestations of pinworms could also weaken his

system, making him liable to pneumonia or inflammations of the stomach, ailments that also could shorten his life. But if he maintains his health, as pandas have a much better chance of doing in the wild away from civilization's germs, he could live for at least eighteen more years.

As for his mother, once her cub has gone for good, she no longer feels concern for it. In the forests she hears the fall courting roar of a male panda and lifts her head to listen. Once more her blood stirs. Once more she starts out on the old mating trek, leaving her scent as she goes.

The panda's year of mating, birth, growth, and separation has been but one in her species's infinite cycle of generations. Each year she will make her arduous journey up and down the mountains, face the harsh climate and the preying dangers that await her along the way, and if need be, fight to protect her own life and that of her cubs. It is a bitter and never-ending struggle, yet the mother panda will always be faithful to her summons. The survival of her ancient race depends on her.

# Epilogue

ACCORDING to Chinese zoologists, the panda's ancestor was a much smaller animal that prospered in the early Pleistocene era. The only fossils of this creature to be discovered to date were found in a cave in Liucheng in the far southern province of Kwangsi.

By the Middle Pleistocene, some 600,000 years ago, the small panda had developed into a much larger animal closely resembling today's giant panda. This panda was widely distributed throughout China; some of its bones have been discovered in the famous caves of Choukoutien outside Peking in northern China. Here they were mingled with those of Peking Man—one of China's Early Paleolithic peoples.

From the Yangtze River southward into Burma the giant panda was a common animal until the Late Pleistocene era. Then man began increasing in numbers, carving out more and more fields for himself from the wilderness. The climate also was changing. The bamboo forests on which the giant panda depended began to dwindle. The animal's range shrank until finally it was reduced to the cold rain forests of the far western mountains.

Europe first came to know about the giant panda in 1869 when a French missionary, Abbé Armand David, who was also a naturalist, went to the mountainous country of Szechuan Province to search for rare plant and animal life. A hunter brought him a dead panda, and he sent the skin and skeleton to a friend in France, naturalist Alphonse Milne-Edwards, for analysis.

Thus an argument started that has been going on for a century. To what family of animals does the giant panda belong?

Everyone agrees that it is a carnivore because, though its main diet is bamboo, it also eats some flesh. The ancestors of the modern carnivores are the miacids, small forest animals that lived some 60 million years ago. About 35 million years ago one branch of the miacids developed longer legs, larger brains, and sharp teeth with which they could tear flesh. This branch is called the canid branch and is the ancestor of the modern dog, fox, and wolf.

About 10 million years later, the canid carnivores began breaking up into branches. One branch developed handlike paws, supple limbs, and blunt teeth that they could chew vegetation as well as flesh. These were the procyonids. Animals such as the raccoon descended from them.

Several million years later, another branch split from the canids; this group was larger, developed heavier skeletons, and had big skulls. They also developed blunt teeth with which to chew vegetation. But their tails were much shorter and their legs heavier. These were the ursids, the bear family's ancestors.

In what family did the panda belong—the procyonids or the ursids? Today most zoologists have solved the question by giving the panda a special genus of its own because it has so many unique characteristics. This genus is called the *Ailuropoda*, which means "cat-footed." Some zoologists place the lesser panda in the giant panda genus. Others believe the lesser panda should remain in the raccoon family, and a very few want to give it a family of its own.

Therefore, the giant panda is scientifically classified as *Ailuropoda melanoleuca*, which means "the black-and-white cat-footed animal." The lesser panda is called *Ailurus fulgens*, or "fire-colored cat." But, of course, most people refer to them as the giant panda and the lesser panda.

The word *panda* comes from the little country of Nepal, which lies south of the Himalayan Mountains. *Panda* means "bamboo eater" and is the native name for the lesser panda, which frequents the forests of Nepal. When Westerners discovered the lesser panda, they adopted the Nepalese name for it. When they found the giant panda farther north, they simply called it the giant panda because it also was a bamboo eater. The Chinese themselves have several names for the panda: white bear, bear-cat, speckled bear, and monk bear. The last name was chosen because the animal likes to wander alone.

Despite all the names the panda has been given, it has never been well known in China. In the seventh century during the reign of the first T'ang Dynasty emperor, "white bear" skins are mentioned as having been among the gifts sent to the ruler of Japan along with two living white bears. But there is no proof that these white bears were pandas. They might have been polar bears obtained from the arctic regions far to the north. Or they might have been the Tibetan blue bear with its straw-colored pelt.

No further mention is made of the "white bear." But during the Ching Dynasty (1644–1911) panda skins were probably among the tributes sent to the emperors from the mountain tribespeople of Szechuan. If so, they were not highly prized by the royal court; the panda pelt was considered much too coarse for royal tastes. So the panda was relatively safe from predators until Western hunting expeditions began coming to the wilderness to bag this strange animal.

The first Western hunters to shoot a panda were the Roosevelt brothers, Theodore and Kermit, sons of American President Theodore Roosevelt. They brought the panda skin and bones back to America, and the animal was mounted and set up in the Field Museum in Chicago. This started the demand for mounted pandas in museums.

In 1936 Ruth Harkness, widow of the wild animal collector William Harvest Harkness, Jr., brought the first live panda to America. The live panda race was on. Soon the Chicago zoo had three live pandas; the Bronx zoo in New York City and the St. Louis zoo in Missouri obtained two each. Over in England the London zoo bought three pandas from a noted wild animal collector named Floyd Tangier Smith.

Since little was known about the care and transportation of captive pandas, many died after their capture, some even before they left Szechuan, others en route to their respective zoos. A large number of those that did survive the trip died prematurely in their new homes.

The two Bronx zoo pandas that had arrived in 1938 and 1939 were both dead by 1941. Zoo authorities immediately sent an order for two more to the West China Union University in Chengtu, capital of Szechuan Province. The university, which had begun to act as a panda clearinghouse, also received an order from Madame Chiang Kai-shek, wife of the Chinese ruler. She wanted to give a panda to the American people for their contributions to United China Relief. To obtain the three pandas the university gathered together seventy hunters who with forty dogs set out to search panda country. Despite their numbers and the large area they covered, they were able to capture only two pandas, which arrived at the Bronx zoo shortly after the bombing of Pearl Harbor.

At the end of World War II, with all three London zoo pandas dead, the British government appealed to the governor of Szechuan Province for two more pandas to take their places. Once again there was a massive search, this time with some two hundred peasants and professional hunters. They scoured panda country for two months before they succeeded in capturing one animal.

The Chinese began to realize that if they had to go to such extraordinary lengths, and take such a long time, to capture one panda, it could only mean that their prized animal, already rare, was dying out. On October 26, 1946, a Shanghai newspaper published this melancholy fact: "Szechuan's precious wildlife, the giant panda, has been hunted to the verge of extinction. This species that has been with us for so many thousands of years is bound to disappear and the world will never see it again."

In 1949, after the establishment of the People's Republic, the new government moved at once to save all its endangered species. The rarest of all, the giant panda, was first on the list. Several regions that the panda was known to frequent in numbers were set aside as preserves. Prohibitions were issued against hunting the panda, not only in these preserves but anywhere else it was to be found. Propaganda teams went to all the remote villages in panda country to explain the new regulations and to enlist the villagers' aid in protecting the panda.

A special bureau was established in Szechuan to advance on-the-spot investigations of the panda's habits in the wild. And though giant pandas can still be seen in zoos, collecting them is now the work of the government. Hunters can no longer go in at will and track them down for profit.

65 : *Epilogue*

Those pandas that are captured are housed at an Animal Husbandry Station from which they are dispensed to zoos around the country. Peking Zoological Gardens got its first panda in 1956. Since then zoo pandas have steadily increased in China. Today there are some eighteen pandas in Chinese zoos. Zoologists at these zoos carry out continuous studies to determine the best methods of caring for the giant panda in captivity.

The panda has become more than a rare zoo animal. Today it is China's ambassador of goodwill, given to selected countries as a diplomatic gesture. No amount of money can buy a panda, but two have been presented to the Moscow zoo, three to zoos in North Korea and two to zoos in France. In 1972 two pandas arrived at Washington, D.C., following former President Nixon's visit to the People's Republic. And in 1975 two more pandas arrived at the zoo in Mexico City.

Meanwhile, in 1963 the first baby panda was born in a zoo—at the Peking Zoological Gardens. A year later a second arrived. Several others have been born since then.

In 1972 and 1973, survey teams of zoologists, making on-the-spot estimates of the numbers of pandas in Wanglang Preserve, determined that at least two hundred pandas are living in this area of 107 square miles. The villagers say this is a slight increase over the past.

But the giant panda's distribution far exceeds Wanglang Preserve. In 1940 a Chinese naturalist named Pen Hung-shou sighted it in Tsinghai Province in the Tibetan highlands. The alpine meadows where he saw it feeding and nursing its two cubs were 175 miles to the west of the nearest known panda country. Further investigation revealed that pandas were apparently not a rare occurrence there. Pen was told in the marketplace of Sining, capital of the province, that a panda skin was rarely seen for sale, not because it was scarce, but because it did not bring as high a price as the pelt of the blue bear.

The giant panda has also been sighted in the mountains of Shensi and Kansu provinces in the north and in those of Yunnan Province to the south. Altogether the area in which it has been seen covers thousands of square miles. No one knows for certain the exact number of pandas in this area, but now that hunting has been restricted, they should be increasing.

Whether or not the giant panda will continue to exist, however, depends on more than man. Several factors have to be taken into account. One is the very specialized diet on which the panda exists. In Wanglang Preserve, where for some twenty-five years no ax has been allowed, the trees have grown tall and luxuriant. If they are left uncut, will they choke out the bamboo thickets beneath them, thus starving out the pandas? On the other hand, will cutting them down to allow new forests to spring up aid or harm the panda? It is a problem that requires careful study, and Chinese zoologists are working on a solution.

Meanwhile, the pandas are safe—at least for the present. And there is good hope for the future now that man has taken an active interest in their preservation. This concern, added to the animal's own stubborn will to survive against incredible odds through the millennia, may preserve its beauty in the bamboo glades for years to come. And the black-and-white form that once awed simple tribespeople will continue to amble shadowlike through sunlit jungles, or frolic under the luminous moonlight—a living heritage from the world's ancient past.

67: *Epilogue*

# Bibliography

BOOKS

ANDREWS, R. C. *Camps and Trails in China*. New York, 1918.

COLLINS, L. R., and PAGE, J. K., JR. *Ling-Ling and Hsing-Hsing*. New York, 1973.

CRESSEY, G. B. *Land of the 500 Million*. New York, 1955.

DAVIS, J. A. *Pandas*. New York, 1973.

FERGUSSON, W. N. *Adventure, Sport and Travel on the Tibetan Steppes*. New York, 1911.

FOX, H. M. *Abbé David's Diary*. Cambridge, Mass., 1949.

GRZIMEK, B. *Grzimek's Animal Life Encyclopaedia*. Vols. 7–13. Zurich, 1968; New York, 1972.

HARKNESS, R. *The Lady and the Panda*. London, 1938.

HEGNER, R. *Parade of the Animal Kingdom*. New York, 1935.

HOSIE, ALEXANDER. *Three Years in Western China*. London, 1890.

*Larousse Encyclopedia of Animal Life*. New York, 1967.

LITTLE, A. J. *Mount Omi and Beyond*. London, 1901.

LU, HSIANG-PIEN. *The Story of the Bear Cat*. Hongkong, 1974. Translated by M. Rau.

MORRIS, R., and MORRIS, D. *Men and Pandas*. London, 1966.

PERRY, RICHARD. *The World of the Giant Panda*. New York, 1969.

ROOSEVELT, T. and ROOSEVELT, K. *Trailing the Giant Panda.* New York, 1929.

SCHAEFER, E. *Ornithologische Ergebnisse Sweier Forschungsreisen nach Tibet.* Berlin, 1938. English excerpts.

SHABAD, T. *China's Changing Map.* New York, 1972.

WALKER, E. P. *Mammals of the World.* Vols. 1 and 2. Maryland, 1968.

WALLACE, H. F. *Big Game of Central and Western China.* London, 1913.

PERIODICALS

*Acta Zoologica Sinica* (Journal of the Peking Zoological Gardens). 1974. Includes the following articles about the panda, translated from the Chinese by M. Rau:

Anonymous. "Investigation into the Giant Panda's Propagation Methods and the Birth and Growth of Its Young."

————. "Rearing of the Giant Panda in the Zoo."

————. "Prevention and Cure of the Giant Panda's Diseases."

————. "Investigations into the Behavior of the Giant Panda in the Wanglang Preserve of Szechuan Province—Material Gathered by Expeditions to the Wanglang Natural Preserve of Szechuan."

————. "A Survey on the Giant Panda."

Chu Ching. "Concerning the Giant Panda's Classification—Peking Institute of Zoology Academia Sinica."

Pei Wen-chung. "Brief History of the Giant Panda."

Wang Tsiang-ke. "On the Taxonomic Status of Species, Geological Distribution and Evolutionary History of the Giant Panda."

Anonymous. "Baby Giant." *Time,* 1936.

————. "Mrs. Harkness and Her Panda." *Time,* 1937.

————. "In Pursuit of the Golden Monkey." *China Reconstructs* (Peking), 1960.

————. "In Search of the Panda." *Discovery,* 1933.

————. "Panda up a Tree." *Animal Kingdom,* 1942.

*69 : Bibliography*

————. "Some Wild Animals in China." *China Reconstructs* (Peking), 1973.

National Zoological Park. "Information on Giant Pandas." Smithsonian Institute, Washington, D.C., 1975.

Roosevelt, K. "The Search for the Giant Panda." *Journal of the American Museum of Natural History*, 1937.

Sage, D. "In Quest of the Giant Panda." *Journal of the American Museum of Natural History*, 1935.

Sheldon, W. G. "Notes on the Giant Panda." *Journal of Mammalogy*, 1937.

Tangier-Smith, F. "Hunting the Panda." *Living Age*, 1937.

Wang Sung and Lu Chang-kun. "The Giant Panda." *China Reconstructs* (Peking), 1973.

Yao Chin-hua. "In the Homeland of the Giant Panda." *China Reconstructs* (Peking), 1974.

# The Gray Kangaroo at Home

# *Prologue*

Australia, about the size and shape of an upside-down United States, is the oldest continent in the world. Washed by successive millennia of rainfall, swept by scouring winds, it has become a land primarily of plains. Only one chain of low mountains, the Great Divide, runs through it from north to south, separating the fringe of wetter eastern lands from the semiarid steppes and vast deserts of the west.

Midway up the Great Divide, in the state of New South Wales, the hundred-mile-long Warrumbungle mountain chain thrusts northwestward. Though the height of the Warrumbungles only ranges from 1,250 to 4,000 feet, their massive buttresses appear enormous, rising abruptly from the plains and culminating in a dramatic tangle of spires, domes, and mesas.

The fantastic shapes of the Warrumbungles are the result of fierce volcanic action which took place some 13 million years ago, when a huge active cone burst upward. The volcano erupted again and again over the next millennia. Its crown finally collapsed, and flows of lava broke from seams in the volcanic walls to radiate in all directions, filling the valleys with basalt.

When the volcano became inactive, it experienced year after

year of erosive wear by wind and rain. The soft sandstone walls under-lying the valleys were eventually worn away, leaving the harder lava to stand alone as ridges, domes, and strangely shaped slabs of upended rock. Out of this wild topography the Castlereagh River flows, circling the mountains on its way to the sea. As it goes it gathers in lesser streams that in times of rainfall dash out of the wilderness to join it.

The craggy wilderness gives rise to a diverse flora. The steep slopes are clothed with many varieties of eucalyptus, a tree unique to Australia. There are approximately six hundred species in the downunder continent, and some eighteen of them are found in the Warrumbungles—among them the ghostly white gum, the gnarled black ironbark, the stately red river gum, the shaggy stringybark, and the wide-branching yellow box. They flower in different seasons of the year, so the honey eaters that frequent the mountains are seldom without nectar.

The eucalyptus shares the forest with the acacia, known in Australia as the wattle because the early pioneers used it to make their first wattle-and-daub shelters. There are some twenty-two species of acacia in the downunder continent, most of which are represented in the Warrumbungles. They grow to different heights and, like the eucalyptus, one or another of them is flowering around the year.

In the valleys and on the lower slopes there are woodlands with large branching apple trees that bear small, hard nuts, not apples, for they are not even distantly related to the orchard variety. They were simply called apples by homesick pioneers because their bark resembles that of the true apple tree.

The dark green cypress pines standing like sentinels among the eucalypti are not members of the cypress family either but simply look like that tree. And the casaurinas, whose drooping pinelike needles fringe the streams and, in the lower valleys, reach heights of 150 feet, are called "she-oaks" because their wood is hard and has a grain similar to that of the genuine oak tree. The she-oak bears small nuts while its counterpart, the male casaurina, is barren.

All these trees add their patterns of leaf and structure to the forests of the Warrumbungles. And despite the fact that most of the trees are eucalypti and acacias, there is still a rich patchwork of colors that vary from airy silver-grays to bright emeralds, shifting with every passage of the wind.

Not all is woodland, however. On rocky, barren red screes the hardy kurrajong flaunts a bright green umbrella of leaves. The wild fig tree flourishes in deep damp gullies and wet crevices, while on the windswept mountain summits, black boys with crooked trunks and great tousled pompoms of grassy hair grab root, and twisted snow gums find anchor.

The Warrumbungle National Park, which has been established here, has not been untouched by the hand of man. Farmers and cattlemen whose holdings skirt the park once tried unsuccessfully to wrest a living from its slopes and plateaus, clearing away here and there stands of trees to make way for paddocks and farming land. During those years the farmers waged a ceaseless war on the kangaroo, which they considered a pest.

But farmers and cattlemen soon found the soil too poor for either farming or grazing. And soon, with the disappearance of the trees and the close cropping of grasslands by sheep and cattle, erosion set in.

Then in 1955 a national park was created from land either donated or purchased by the government in the most spectacular part of the Warrumbungles. Gradually the area of the park expanded, and today it covers some 62,000 acres. Camps for visitors have been provided and nature trails have been built. But ample slopes and plateaus have also been designated restricted areas to provide a sheltered retreat for wild life.

The eastern gray kangaroo, one of the largest of the species of kangaroo, has lived in these Warrumbungles for thousands of centuries. Also called the forester kangaroo because it is found primarily in the forested regions of eastern Australia, it now roams freely through the park lands and forests, protected from its most dangerous predator—man.

It shares its home with other species of kangaroo—the wallaroo and three varieties of wallaby—as well as smaller furry marsupials or pouched creatures. Overhead the trees are noisy with a rich community of birds—parrots and honey eaters and kingfishers, many not seen elsewhere in the world.

Among such company the forester begins its life as a tiny grublike embryo and lives out its web of days.

# ONE

# A Joey Is Born

It is late September, springtime in the Warrumbungle mountains of the downunder land. Sunlight filtering through the flowering eucalyptus woods flecks the ground with shadows. Where the shade is deepest, a forester kangaroo doe and her one-year-old young sit together. Both wear coats of grayish brown fur shading to a lighter color across their chests and abdomens. The tips of their tails and the fingers of their hands are almost black, giving them an elegant gloved appearance.

The doe would be about four feet tall if she were standing upright, which she sometimes does, balancing herself easily on her ten-inch-long feet and a muscular tail almost the length of her body. Her feet and tail act as a perfect tripod whenever she rises to her toes.

Her haunches are large and muscular, but her long legs are thin and bony and her chest narrow. Her slender arms seem underdeveloped in contrast to her thighs. Across her lower abdomen stretches her furry pouch, a fleshy envelope with elastic muscles along its lip that hold it firmly against her body.

For some three hundred days the pouch has been home to the young kangaroo or joey as it is called in Australia. Though it has now left the pouch permanently, it still needs its mother's milk supply to augment the grasses which it is learning to eat. It will be eighteen months old before it is completely independent. Until then it must remain close to the doe or it will die. During this stage of its development, the joey is also known as a young-at-foot.

The joey, which is only a foot and a half tall, is a female and an exact replica of the doe, even to the tiny pouch concealed in fur. Both joey and mother have long pointed muzzles and large dark eyes fringed with heavy lashes. There is an alertness in the cock of their pointed ears which stand upright, swiveling to catch every forest sound.

But there is another kind of expectancy about the doe today. She is not occupied with her usual pursuits of feeding and resting. Instead she is spending most of her time cleaning her expanse of chest and abdomen, using her delicate smooth tongue and the long sharp claws on her fingers. Of particular concern to her is her pouch.

Every so often she leans forward and pulls the lip of the pouch sideways with her forepaws so she can slip her muzzle inside. She licks and nips at the lining, which became covered with dark brown scales after she turned her young out of it several weeks before. She keeps at her work even though she has long since licked and nipped away all the scales, leaving the pouch meticulously clean and very soft.

Inside the pouch there are four teats ranged in two rows along the doe's abdomen. The one on the upper-left-hand side is the only one on which her joey has suckled. Now it is elongated to some two inches and protrudes over the lip of the pouch. The other three teats, which are very small in comparison, are tipped with budlike knobs in preparation for the advent of new life that began a little more than a month before, when the doe mated with a buck kangaroo. It was just before this yearly mating that the doe evicted the joey from her pouch.

Though the joey no longer has access to the pouch, the bond between the doe and her daughter is still very close. Sitting beside her mother, the joey copies all her grooming movements. She mimics everything her mother does because these behavior patterns do not come by instinct but must be learned through imitation.

The day nears its end. Shadows deepen in the glade, though sunlight still lingers in the tops of the trees, where the noisy friars sip nectar through long beaks. They are clumsy bluish gray birds with heavy wings and naked black reptilian heads, and they squawk and squabble endlessly among themselves.

As dusk comes, the clamor of the birds quiets except for the curlew, which now finds voice. Its eerie whistle *wee-lo wee-lo* rises and falls upon the hushed air. It is now time for the creatures of the twilight and the night, as are most Australian animals.

A plump gray koala that has been asleep a good part of the day in the high fork of a eucalyptus stirs itself, filling the air at intervals with a loud coughing mating call that always fades away into throaty mumblings. It will spend most of the night feeding on the leaves of the eucalyptus.

Brushtailed possums venture from their nests in hollow tree trunks and branches to feed upon flowers and green shoots, every female carrying a young one on her back or in her pouch. Their bursts of chatter punctuate the hush of falling twilight.

Sugar gliders looking much like small flying squirrels also come out with the dusk. Launching themselves airily from high branches, they sail down through the air on webbed membranes, landing on the ground with a soft *clup clup*. After a swoop of some fifty yards, they will immediately scurry up another tree to rob it of its blossoms before swooping down again.

The sugar gliders congregate in family groups. Now and then an altercation among them disrupts the peaceful night with a bout of angry screams which fade away to peevish gruntings.

In the lower thickets the perky pigmy glider, no larger than a mouse, flits here and there on tiny membrane sails. Its long feather-like tail serves as a rudder by which it steers itself on its short flights in search of insects.

High in the trees clouds of bats hover, whispering a soft *tsk tsk* as they garner insects, while on the forest floor tiny furry creatures bustle out from their nests under rocks and in underground burrows to prey on insects and each other.

Night has its own strange sounds bursting out above the steady thrumming of the crickets. The melancholy *mobook mobook* of the stocky boobook owl floats through the trees. The shrill cries of owlet nightjars are punctuated occasionally by the deep *oom oom* of the frogmouth owl, who perches so quietly on an old tree stump that he is indistinguishable from it. His beak is as wide and broad as a frog's mouth and he is a fierce hunter of moths.

The doe and her joey remain unaffected by the approaching night, but when a terrible scream suddenly echoes through the darkness like the sound of a child being strangled, the ears of the doe jerk upright, twitching, and the joey begins to shiver. Then they relax as the sound quickly falls away. It was only the cry of a barn owl.

The night stretches overhead with a galaxy of stars. Here in the downunder world, Orion stands on his head while the great Southern Cross wheels through the sky as though drawn by its pointer stars. With every passing hour the doe steps up the pace and frequency of her grooming—pouch, chest, forearms, legs. But she gives special attention to her abdomen, from the lip of the pouch downward to the birth passage in her genitals, which she also carefully cleans.

Night fades, and with the first crystal light of dawn, birds break into a jubilant chorus. The currawongs, a species of magpie, converse with each other over and over in a duet which gives them their name. *"Curra"* cries one. *"Wong,"* replies the other. *"Curra-wong, curra-wong."* And the great black ravens solemnly enunciate their sepulchral *au au auuuoh.*

The doe takes no notice. She is busy excavating a saucer-shaped hollow in the soft earth. When it is to her liking, she squats in it with her tail stretched out behind her. Her legs are thrust forward, the toes of her large feet point skyward, the heels push against the rim of the hollow. Bending her upper body forward while supporting herself on heels and tail, she concentrates on the birth passage between her legs, licking and nipping at it furiously.

After their first outburst the birds have fallen silent. Then as the sun brushes the tops of the trees, the gray and pink galahs and the white yellow-crested cockatoos churn up the day with their clatter and bustle. The nocturnal creatures, shy of light, begin their retreat to shelter.

Only a month after her mating the kangaroo doe is in the throes of birth. A small amount of thick, yellowish fluid mingled with a little blood issues from the birth passage and is quickly licked up by the doe. It is followed by a round sac less than an inch in diameter, which contains the yellowish wastes of the tiny embryo.

Immediately afterwards the embryo itself appears head first in its own fluid-filled envelope. The kangaroo rarely has twins, and if she does, only one could survive. Two young in the pouch would destroy each other.

This tiny creature, which is about three quarters of an inch long, is hairless and weighs scarcely a thirtieth of an ounce. Prominent blood vessels show clearly through its thin pink skin. Its head is large, its undeveloped eyes and ears are still covered with a membrane. Its nostrils however are flaring, and its sense of smell

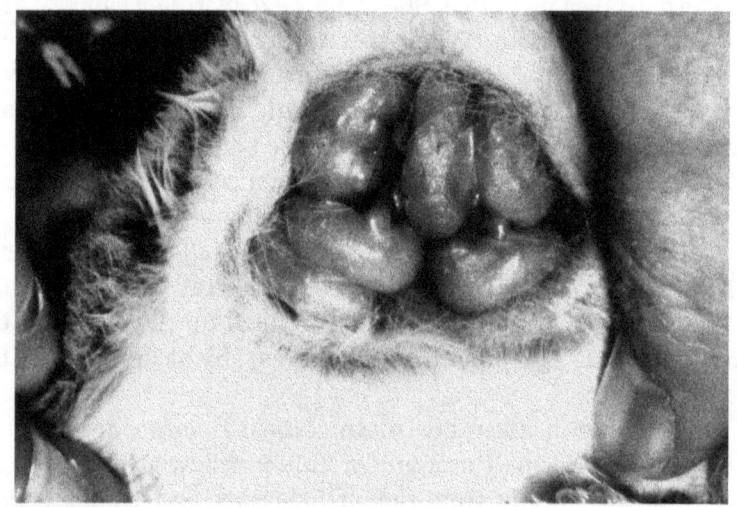

A kangaroo seldom has more than one joey.  Smaller marsupials,
like those babies pictured above, have anywhere from six to ten tiny embryos.

is probably already keen. A large tongue protrudes from its circular mouth.

The muscles of the tiny creature's forelimbs, shoulders, and neck are all well developed. And the fingers of its well-formed hands are armed with sharp, curved claws. But its hind legs and tail are only buds. It seems unbelievable that this grublike creature will ever grow to be as big as its mother—if it lives.

Still enveloped in the sac, the embryo twists itself quickly around so that it is on its stomach with its head pointed toward its mother's pouch. Shortly afterwards the sac which holds it is ruptured either by the sharp claws of the tiny creature or by the helpful licking of its mother or both. The water leaks out and the embryo is ready to start its long, hazardous journey to the pouch rim some six inches away.

Sharp claws catch hold of the hair at the birth opening, and the embryo begins pulling itself upward, moving its head from side to side as though sniffing for the pouch. The umbilical cord to which it is still attached stretches until it is taut. Far from its goal, the embryo finds itself anchored. It meanders back and forth across its mother's abdomen, tugging and straining to pull free. Finally, perhaps aided by the doe's licking, the cord snaps and the embryo can continue its journey.

It will get no more help from its mother, but must make the rest of the journey alone, its first test of fitness. Weaker embryos may be so slow that they die of exposure before they reach the pouch. Others may lose their sense of direction and wander off in the wrong direction, finally losing their grip on the mother's fur and dropping off. If this happens, the mother cannot help them in any way. Still others may be dislodged if the mother, suddenly frightened, bounds away in a panic. Only the healthiest embryo can survive the perilous trek.

ILLUSTRATION                                    : 85 :
Below: embryo climbing to mother's pouch
Above: embryo suckling teat in pouch

Painfully, the little embryo makes its way through the thick soft fur of its mother's abdomen. Hand over hand over hand, it hauls itself up. Its mother watches its progress, but she cannot use even her sensitive tongue or lips to shove it along. The gentlest touch at this time would kill it.

Several times the embryo wavers uncertainly. Then it doggedly continues on. After three long minutes it reaches the lip of the pouch and enters it. But its fight for survival is not yet over. Once inside it must locate a teat. If it does not locate it soon, the embryo will die inside the pouch.

Pressed between the wall of the pouch and the abdomen, the embryo gropes with its mouth and large tongue. It avoids the long teat on which the young-at-foot is still suckling. Finally it makes contact with the teat on the upper-right-hand side of the pouch. As the embryo's mouth closes over the bead at the tip of the teat, it begins to swell until it fills the whole mouth. The tiny creature starts to suckle.

Teat and embryo are now so firmly attached to each other that a careless attempt to pull them apart would rupture the embryo's mouth and kill it. This was why the early settlers in Australia believed that the kangaroo's joey was actually born from the bud on the end of the teat.

Not long after the embryo enters the pouch, the placenta which cushioned it in the doe's womb is ejected. She licks it up and then continues to clean herself of the yellowish fluids that are still oozing from the birth passage. She also licks clean the wet trail the embryo has made through her fur. Occasionally she licks the rim and the interior of the pouch along with the embryo inside. While all this is happening, she does not forget her young-at-foot. When it nuzzles her she leans over and waits patiently while it suckles on the protruding teat.

Kangaroo milk differs a great deal from cow's milk. The content of the milk the joey is drinking from the elongated teat is different from the milk in the teat which the embryo is suckling. Both contain the same amount of sugar, but the large teat has been providing a greater and greater amount of fat and protein over the months to satisfy the needs of the growing joey. On the other hand, the milk which the embryo is drinking contains about five times more iron and three times more copper, since the embryo lacks these minerals at birth.

Toward midday the birth fluids stop flowing. The doe wanders with her young-at-foot away from the shallow depression where the birthing took place. She finds a shady spot under some trees and stretches out to rest. In the months ahead she will have to eat a great deal if she is to provide enough rich milk for both the tiny creature in her pouch and the young-at-foot. Her appetite will become enormous. But now she is only tired.

# TWO

# *Downunder Spring*

In midafternoon the kangaroo doe and her joey begin hopping randomly through the forests. Even when the doe stands upright, it is impossible to distinguish the embryo in her pouch, and she herself ignores it. Doe and joey move at a leisurely pace, stopping now and then to feed.

Side by side, mother and daughter browse with both hands and hind feet flat on the ground and their heads swinging from side to side. When they want to move forward to a fresh grazing patch, they plant their hands in front of them. Then, resting on their front feet and using their muscular tails as props, they give a little hop that brings both hind feet level with their hands. To cover any distance they repeat this maneuver over and over.

Though foresters spend most of their time under the trees, they are not very interested in forest foliage. Their diet is made up of grasses, preferably the wiry kind as long as it is short. Most of their browsing is done in open meadows.

The teeth of the kangaroo, like those of all herbivores, are adapted to grazing. Both the upper and lower jaws contain four, sometimes five, cheek teeth on either side, all of which have erupted by the time the kangaroo is six years old. There are three sharp front teeth or incisors in the upper jaw. The lower jaw has only two incisors, an inch and a half long each. Between the molars and the front teeth there is a two-inch gap.

What makes the kangaroo's lower jaw different from those of other herbivores is that it is separated into two halves which can move somewhat independently of each other. When the kangaroo opens its mouth to grasp the grass stems, the lower jaw moves forward and downward at the same time. The mouth closes on the grass stems, which are caught between the upper and lower incisors. An upward jerk of the head snaps off the grass stems. Then the tongue and lips arrange the stems in the gap between the front and back teeth and feed the ends to the molars. Here the masticating is done by sideward movements.

Constant chewing of abrasive foliage grinds down the molars. As the front ones become too worn for use, they fall out and are replaced by the ones behind, the whole row of teeth simply moving forward in the jaw. The kangaroo has only one set of molars, so if it grows old enough to lose most of its cheek teeth, it will die of starvation.

Like all kangaroos, the doe rests a great deal between browsings, but she spends very little time actually napping. Occasionally while relaxing she will bring up a portion of her food as cows do, but it is much harder for her. She has to hold her head back and cough to get the wad up. Green saliva trickles down the corners of her mouth as she succeeds. Sometimes she chews the wad. At other times she simply holds it in her mouth.

The little joey at her side mimics her coughing and gagging until it too brings up a tiny green globule. But since she doesn't know what to do with it, she lets it fall to the ground.

The glory of the downunder spring envelops mother and daughter as they wander through it. Overhead the eucalypti are flowering. Banks of acacia are thick with golden blooms mingled with the purple blossoms of the climbing wild pea. In shady places delicate wild iris make splashes of lavender, while on more open sandy slopes pink boronias and soft white flannel flowers, red fuchias and golden everlasting daisies spread gay shawls of color. The ghostly white gums rise among them, their pale trunks scribbled over with the hieroglyphs of tiny insect borers, ribbons of old bark hanging from the branched forks.

Already the rainbow bee eaters are flashing through the forest or sitting motionless on twigs, glittering like enameled birds. They are migrants from the subtropics of northern Australia and have traveled southward to the Warrumbungles. Here they will dig three-foot-long curved tunnels in eroded banks or sand ridges, lay their eggs, and raise their young, safe from most predators.

Pausing to drink at a mountain stream, the doe and her joey startle another tunneling bird, the tiny freckled pardalote. It darts on miniature wings like a bullet from its hole in the muddy bank and is off to forage for insects.

Yellow robins little larger than sparrows but with bright yellow breasts perch sideways on eucalyptus trunks, prying away the loose bark to get at the insects beneath. The little birds' tiny cup-shaped nests concealed in thick scrub are covered outside with dainty mosaics of lichen peeled from rocks. Soon each will hold three or four greenish eggs.

Honey eaters large and small flock about the flowers in the tops of the trees. Blue and red mistletoe birds congregate round the yellow-green clumps of parasitic mistletoe that dangle from the branches of high eucalypti. The little bird's chief food is the mistletoe berry, and it plants its own crop simply by sitting sideways instead of crosswise on a branch. This causes its droppings, which are full of mistletoe seeds, to land on a branch, where they will take root, rather than on the ground, where they will die.

Turquoise parrots and crimson rosellas have hidden nests in

hollow trees and fence posts. And in a wide expanse of open flat land, the doe and joey come upon a strutting emu. The huge flightless bird, standing some five feet tall, paces solemnly along on its large horny three-clawed feet, holding high its long naked neck and head. It is armed with a strong, sharp bill, and as it struts by, it eyes mother and joey coldly through glassy red eyes, a hollow *boom boom boom* sounding from deep within its throat.

The joey, no longer able to hide in her mother's pouch, is apprehensive of every strange sight and sound in the wilderness. At the emu's *boom boom boom* she crowds against her mother, shivering fearfully. But it is only the female's mating call, produced by expelling air from an inflatable sac just beneath the skin covering the windpipe. With a flurry of tail plumes the emu passes on. Her mate has the task of preparing the nest—a shallow depression in the ground which he will line with grass, leaves, and small sticks. Here the hen will lay a dozen or more five-inch-long, dark green eggs. It will be the male's chore both to brood the eggs and to guard the striped chocolate-and-straw-colored chicks once they are hatched.

With the emu on his way the joey relaxes, but not for long. Even harmless lizards startle her. The warming days of spring have been wakening them, and they are coming out of their partial hibernation to lie about on rocks and logs, sunning themselves. To expose as much surface as possible to the sun, they flatten their bodies. As they absorb the warmth, they become more active, but they will have to bask many times during the day to restore their body heat, for their metabolic rate is slow.

Each type of lizard has its own form of protection. The little skinks run like lightning when they are pursued, zigzagging and turning back on their tracks as they scurry for cover. Some of the larger skinks have more novel methods of driving off attackers. When the joey comes too close to a blue-tongued skink, pop goes

A snake comes to life in the sun

Goana or monitor lizard

Shingle-backed Lizard

its mouth in a wide gaping cavern and out flies a fat, bright blue tongue. A loud hiss emerges from the throat as the body inflates. And away goes the joey in a flurry of alarm.

She gets another scare from a two-foot-long bearded dragon that lies basking in the sun, so motionless it is almost invisible. The joey just misses hopping on it, whereupon it too opens a cavernous mouth, revealing a bright yellow interior. At the same time a spiky beard unfolds from its chin and bristles aggressively. The joey doesn't need to be frightened off by the short lunge and snap of the powerful jaws. She has already bounded away.

The most formidable of the lizards, the six-foot-long monitor or goanna, as it popularly called, is a common sight in the Warrumbungles. Too large for hibernation, the goanna has passed through a lean winter trying to keep warm while feeding on the meager insects and small mammals it can find. Now it comes slithering on huge clawed feet through a patch of high grass. Aware of the joey nearby, it rears up above the grass for a closer look at what might possibly be a satisfying meal.

But the joey is much too large for the goanna, which quickly loses interest. Long before the joey has made it back to her mother, the lizard is slithering to a tall eucalyptus with a hollow high in the trunk. Hopefully there will be possums in the hollow. The goanna begins to climb upward, its large claws making loud scratching sounds on the loose bark.

Suddenly there's a chittering and a wailing from deep in the hollow high above. The possums have recognized those scraping sounds and are hysterical with fear. But they make no attempt to run away, and soon the goanna, well fed and drowsy, is basking in the sun.

# THREE

# *The Gathering of the Mob*

The doe and her joey spend the night grazing and resting in a leisurely fashion, but when dawn brightens the sky, the doe seems seized with new purpose. She moves at a faster rate, taking hops that cover three to four feet at a time. When she outdistances the joey, she stops and clucks impatiently at her, pausing now and then to let her catch up.

Finally the two bound through gnarled old apple and ironbark eucalypti and enter a wide, grassy flat at the foot of which tall, drooping casaurinas fringe a quiet stream. Several other kangaroos are already scattered about the clearing, grazing quietly. They are not all foresters. Two red-necked wallabies are browsing in one corner. They resemble the larger foresters in appearance; the only difference is their smaller size and the reddish tinge of fur across their shoulders.

At the edges where the clearing melts into brush, several wallaroos or hill kangaroos, who make their home in rocky terrain, have come down to feed and drink. With their shaggy, brownish black coats and unkempt, hairy faces, the male wallaroos look like stooped old men. But their appearance is deceptive, for they are tough and muscular animals. The soles of their short, broad feet are covered with thick rough pads that enable the wallaroos to keep their footing on the great slabs of tilted rock which they mount in a series of powerful vertical bounds. The little wallaroo does in their pale gray fur look small and delicate beside them.

The wallaroos are not sociable creatures. Unless they are mating, males and females tend to gather in separate groups or lead solitary lives. As soon as they have fed and drunk, they will retreat to the high rocky outcrops that are their haunts.

There are more foresters in the clearing than either of the other two species which they ignore. They brouse singly or bunched together in groups of two or three. Such a gathering of kangaroos is known as a mob. In choice grazing sites all through the Warrumbungles, mobs like this one are meeting.

Most of the groups that make up a mob have only two members, a doe and her joey. When an embryo is in the doe's pouch, the family unit becomes three. It also becomes three when the female enters her mating season and attracts a male to her, but this lasts only a brief period, since there are no permanent attachments between males and females among kangaroos.

Wherever there are good grazing grounds, the size of the mob increases, but it never becomes a tightly knit unit which acts together. Kangaroos are too independent for this. At any time a single animal or one of the small groups may leave one mob and join another or form a small mob of its own. Mobs are always changing size and composition in this way. Small mobs may also gather together to form a larger one, and large mobs often split up into smaller ones. The only constants drawing them together are good communal feeding grounds and their preference for company. Despite their independent spirit, kangaroos are a gregarious lot.

This particular mob has six does with their joeys. Half the joeys have left the pouch permanently because the mothers are either about to give birth or are carrying embryos in their pouches.

Two immature males of about the same size as the does are grazing together. From a distance it is difficult to tell the does apart from the young bucks, which are not yet three years old, the usual age to begin mating. A few bucks are able to mate earlier, while others are not ready until their sixth year.

Three mature bucks are feeding singly. They are much larger than the does; the biggest of them stands some seven feet tall and weighs more than 120 pounds. Mature bucks are called "boomers" in Australia because of their habit of signaling danger with hard thumps of their heavy feet as they bound away.

All three bucks have massive muscular chests and powerful arms, but none of them carries excess weight. Rather, there's a lean, bony look about them.

As each doe enters the clearing, the bucks lift their heads and sniff. If they were to pick up the scent of mating time from a doe, they would amble to her and with nose and tongue try to rouse her mating urge. But since none of the does gives off the telltale odor, the males return quietly to their browsing, moving methodically forward, leaving a yard-wide swath of cropped grass behind them. Occasionally, however, one of the mob will lift its head, look around, and decide there's a better grazing ground a dozen or more yards further on. With a bound it's off to browse on the new site.

There is very little verbal exchange between the grazing kangaroos. Their repertoire of sounds is limited to coughs and grunts, sucking and clucking noises by the adults, and bleating by the joeys. Most of their communication is done by sniffing. They may merely sniff in the direction of another kangaroo, or they may greet each other at close range with an exchange of sniffs at the nose, the mouth, the pouch, and the base of the tail. Then, after every sniffing, they return to their grazing.

Suddenly the peaceful scene is broken by the two younger bucks who without warning rise to their hind feet and confront each other. Coughing sharply, they interlock hands and arms and begin to circle round and round.

Finally they come to grips, embracing each other with powerful forearms. Now there's a tug-o-war to see which can topple the other. Neither succeeds, so dropping the embrace, they begin striking out with sharply-clawed hands that reach for eyes and ears. To protect these sensitive areas, each kangaroo holds his head high, exposing his throat to the other's raking claws.

But the two are well matched; neither scores a victory, and they move on to a kicking contest. Balancing themselves on their tails, they aim at each other's exposed bellies using both feet at once.

Despite their apparent struggle, the bucks are really only practicing for the time when they may have to fight in earnest for the privilege of mating with a doe. But for the does and joeys, it is good entertainment. They stop their grazing and sit upright to watch the contest, which comes quickly to an end when one young buck topples his opponent.

As the loser creeps off on all fours, tail dragging behind, an excited thirteen-month-old male joey rushes up to challenge him with a flurry of small, clawed hands. The defeated buck relieves some of his frustration by delivering a sharp cuff that sends the joey bowling end over end. Bleating with surprise and pain, he hops back to his mother for a comforting guzzle of milk.

As the sun rises in the sky, the kangaroos begin their early morning grooming. Even the joeys join in, though in a haphazard manner, so that after they have finished, their mothers give them a second going-over. But it is as much to show affection as for cleanliness.

Grooming is a painstaking ritual. Teeth and clawed hands bite and scratch out the ticks, mites, and fleas that constantly torment the kangaroos. Then the tongue works over abdomen, thighs, chest, and arms. The forearms are licked and wiped over the face to wash it, in the manner of a cat. Finally the clawed hands rake through the coat, carefully arranging the hairs in orderly rows.

The kangaroo uses the toenails of its large hind feet to clean the hard-to-reach places. It has only four toes on each foot; the center toe is the longest and largest, sheathed with a sharp wedge-shaped nail. On the outside of each foot, another smaller toe with a similar nail flanks the large toe. On the inside there are two very small toes bound together by skin; both these two toes have curved claws.

The coupled toes are useless as far as getting around is concerned, never even touching the ground. They only serve as the kangaroo's grooming comb and are especially adapted to such delicate work as cleaning out the corners of the eyes and inside the ears. After each use the kangaroo carefully cleans the crevice between the claws by running one of its sharp incisor teeth through it.

After grooming the kangaroos return to their grazing. But the day, which began quite cool, is heating up to seventy-five degrees Fahrenheit. Kangaroos cannot stand too much heat. They have sweat glands, but, like many other animals, can't cool off by perspiring alone. Instead they pant, but only through their noses. Panting causes evaporation, which cools them. The hotter they become, the harder they pant, though this makes their bodies lose a great deal of water. To conserve energy kangaroos seek out shade when the day grows warm.

The three mature bucks are the first to move under the trees. They are larger and the heat affects them sooner. The immature bucks and the does go on grazing a while longer. Growing kangaroos eat fifty percent more than adults, and the does their size eat even more to ensure a good flow of milk for the large and hungry young-at-foot.

While their mothers graze, the joeys spend most of their time playing. They practice leaps and engage in miniature dueling

Siesta time-The doe in the background is suckling her half-grown joey

matches with bushes and tree stumps. One more daring than the others goes up to an adult male and begins throwing light punches at his massive haunches until it is cuffed away.

Finally does and joeys are also ready to relax in the shade. The clearing is now free of kangaroos. It would take sharp eyes to see their forms, so well camouflaged are they in the grayish shadows under the trees.

But even when they are resting, the kangaroos are seldom completely quiet. They sleep in snatches if at all. Their eyes are usually only half-closed, and their ears are always twitching to catch any signal of danger. Now and then one of them will rise to scan the area before stretching out again.

Much of their time is spent licking down the fur on the chest and forearms, wrists and genitals. There is some rolling around on the ground and almost constant scratching to get rid of the persistent parasites which still annoy them. Occasionally the quiet is broken by a loud, rasping sound as a male sits upright and rakes his claws across his ribs. Fortunately the kangaroo has a tough hide.

As the sun rotates through the sky, the shadier places grow sunnier and the less-favored spots fall into deeper shade. It is then that the adult males show their dominance. The largest buck gets up languidly and approaches the doe which is now lying in the best shade. Coming up from behind he grasps her around the middle with a sharp exhalation of air—*ha!* She understands and quietly relinquishes her position to him. He flops to the ground and sends dust flying as he kicks away the warmed layer of upper earth, pushing himself into the cooler ground below.

The second adult male just makes a grab at one of the juveniles whose space he covets. The young buck also gives way without a fuss. The third adult buck hops over to the second juvenile and stands over him in fighting stance, grunting. The space is quickly vacated. The round of musical chairs is played again and again as the sun changes its position.

In the late afternoon a cool breeze springs up, and the kangaroos stir themselves to return to their grazing, occasionally shuffling down to the stream to drink. The nursing mothers especially feel a need for large quantities of water.

The mob feeds long into the dusk. Then singly or in groups the kangaroos fade into the dark forest. The doe and her joey are alone again. Throughout the night they will drink frequently as they browse. But early morning will find them bounding back again to join the mob in the clearing.

# FOUR

# Forest Neighbors

As spring moves on to summer, the rains increase in number and size. Heavy-bellied clouds drift southward from the subtropics of northern Queensland, where the monsoons are in progress. With a crackling of forked lightning and a boom of thunder, they dump their wet loads on the Warrumbungles.

Normally when the skies are overcast, the kangaroos become more active, browsing throughout the day. If it drizzles they simply turn their backs to it. But they seem to know when a really big rain is on the way. Just before it comes, a kind of excitement runs through the mob. The kangaroos bound here and there aimlessly. When the downpour arrives, they make for the forest and shelter.

Then everything takes on a sodden look. Eucalypti and acacias droop under the weight of water. The leathery leaves of the wild fig tree loose miniature cascades with every gust of air. Water fills the streams until they overflow their banks and rush in wide swaths down the steep mountain slopes.

In the deep, damp gullies the miniature jungles of fern and tangled clematis and wongawonga vines are drowned in an avalanche of water, and the shaggy, somber-coated swamp wallaby that prefers wet places is forced to higher ground.

But suddenly the skies clear and the woods undergo a dramatic change. Foliage sparkles with light and trees steam with a rich eucalyptus fragrance. The streams still rush along, though some of them have been half-silted up with loose sand and boulders washed from the stripped hill slopes. Elsewhere the water swiftly disappears into the sandy soil.

The raucous laughter of the kookaburra, a species of king-fisher, echoes triumphantly through the forest. Going down to drink at a swollen stream, the doe and joey surprise one of the large gray-and-white birds knocking the life out of the cicada it has just caught. Cicadas and grasshoppers are good fare for the kookaburra young nesting in the hollow trunk nearby.

Sea eagles circle overhead to take advantage of the life in the stream, and a large gray water rat searches for worms and shellfish in the mud. His flat, sleek head, very much like that of a beaver, delves into underwater crevices, and presently he rises to the sur-face with a mussel clutched between his front paws. Turning over on his back, he floats in the water, playing with the mussel on his chest. Then tiring of the game, he gives the tightly-closed shell a nip just in the right place with his sharp teeth, and it springs open. He eats the contents, tosses the shell away, and clambers out on a log to fluff up his thick fur.

The gray kangaroo is a creature of the dry forest, and when rains cover the mountain slopes and valleys with lush green grass, there are occasional outbreaks of a parasitic disease called coccidio-sis. Then death sweeps through the mobs with devastating sudden-ness. More rarely there may be an epidemic of lumpy jaw, which causes a cancerlike growth of bone on the jaws, immobilizing them so that the animal dies of starvation.

Death from disease is always a threat to the little young-at-foot. But there are other hazards as well to threaten it. Its life will be in jeopardy on many occasions before it reaches full independence at eighteen months. Only half the joey population will survive this period of testing.

There are any number of simple accidents which can cut a joey's life short. Still dependent on their mothers' milk, many die when they become accidentally separated from the does. Some fall into narrow crevices from which they cannot extricate themselves and starve to death. Just practicing her leaping skills is dangerous for the joey because kangaroo bones are among the hardest and most brittle in the world. The whole weight of the body placed on one leg can snap it. This happens occasionally even to adult kangaroos; unable to do more than crawl, they cannot obtain sufficient fodder and die a slow death.

But whatever dangers may lurk in the outside world, the embryo inside the doe's pouch is safe as long as its mother is strong and well and it is not dislodged from the teat. Sudden changes in the weather will not disturb it. Humidity and temperature in the pouch are carefully regulated, protecting the little creature until it has developed sufficiently to control its own bodily functions.

The embryo is growing steadily. By the time it is two weeks old, its female sex is apparent. Within twenty-two days its tail has enlarged from a mere bud to an inch in length. Its legs and feet are elongating at almost the same rate. Now its presence is revealed by a slight bulge in the doe's pouch which, week by week, will become larger.

Every day the doe licks the interior of the pouch and the tiny joey, keeping both clean and groomed. But she will not develop the same close affection for the embryo that she shows her older joey. Only after some six months of pouch life, when the little creature finally thrusts its furry head above the rim of the pouch, will she exhibit any real fondness for it.

The slow maturing of the forester young is almost equaled by the single young of the koala born a few weeks after the kangaroo embryo and now tucked away in a pouch which opens backward. The koala young will be eight months old before it outgrows the pouch and begins riding about on its mother's back.

The smaller the marsupial is, the larger seems to be its brood. The tiny feathertail glider, no bigger than a mouse, gives birth to eight young. But since she has only four teats half her newborn die.

Another tiny marsupial, the yellow-footed marsupial mouse, has an even larger litter. Flushed out of a tussock of grass by the grazing forester and her joey, she races off on yellowish red feet in a flurry of grayish yellow fur. Twelve young sway like peanuts from her teats, which are set in a shallow, open pouch.

Brilliant butterflies and large moths are about in numbers, and the air is filled with the hum of bees. Most are the aggressive Italian variety imported in recent years, but there are also a few swarms of the native Australian bee, a small black insect no larger than a housefly.

The Australian bees are stingless, but it's a different matter with other insects, as the little joey discovers. Grazing one day beside her mother, she chances to nibble the grass over a nest of green-headed ants and is repaid with a sharp sting on the nose. Bleating with pain, she turns to her mother as always for a comforting sip of milk.

But another creature that just then comes trundling out of the woods does not share the joey's distaste for ants. She is a strange, squat little animal, dark brown in color and covered with quills which have developed from greatly enlarged hairs. She lumbers over to the green ants' nest and examines it carefully with her long beak-shaped muzzle from which a sticky tongue darts in and out.

This is the spiny anteater, the echidna. She has just come out of hibernation and will soon be mating, after which she will produce an egg. This she will place in her pouch, which is actually just a muscular depression sunk in her abdomen and formed only at breeding time. Out of the egg will eventually emerge a naked young anteater. Until it grows its quills and becomes too uncomfortable for its mother to carry around, it will live in the open pouch, lapping up its mother's thick, rich milk, which does not flow from teats but from milk glands under the skin.

In order to produce an egg and, afterwards, the milk to sustain her young once it is hatched, the echidna needs a great deal of fat to build up her energy. The green ants can scarcely satisfy her, and she continues on until she comes to a large mound covered over with a roof of tiny hard grains of sand cemented together by rain and sun. Inside in winding galleries and small dark chambers lives a colony of meat ants.

At this time of year the virgin queens of the meat ants have not yet left the hill, and their bodies contain forty-two percent fat. This is just what the anteater needs. Choosing the warm northern side of the mound where she knows the ants always congregate during the cooler months of the year, the anteater hunches her shoulders and begins to burrow with her bulldozerlike strong front claws.

Kangaroos have a great deal of curiosity about strange things, and once the joey realizes the anteater is no threat, she follows the creature to see what she is up to. The joey sneezes and hops back a little as dust flies from the anteater's claws.

Then the mound is opened and the anteater begins greedily lapping up ants of all sizes, along with a quantity of dirt. She has no teeth, only horny ridges on the back of her tongue and her palate. These work together to grind up the food. The dirt which is swallowed aids digestion in the stomach, which is rather like a gizzard.

The joey's interest flags as she watches the anteater methodically rooting through the mound. When she hears her mother's admonitory *cluck cluck cluck*, she hurries back. The doe does not like to have her offspring out of sight for long, and when she feels the joey has strayed too far, she will always call it back with that warning clucking sound.

# FIVE

# A Tragic Flight

Even in the peaceful Warrumbungles there are dangerous predators. Spring has been wakening the snakes from their winter's hibernation, last year's skins hanging in loose wrinkles on their ropelike bodies. Soon the snakes will be sloughing them. Starved and irritable from their winter's fast, they are not to be trifled with. The tiger snake and the brown snake, which live in the Warrumbungles, are among the most poisonous in the world, but even the tiny white-faced witch snake that lives under boulders and feeds on small mammals has a bite as painful as a hornet sting.

None of these snakes troubles the doe or her joey. But one day, bounding ahead of her mother, the joey surprises an eight-foot-long diamond python sunning itself on an open patch of ground, its dark green, patterned coat serving as effective camouflage.

The python, whose nests are usually in rocky areas, is a silent harbinger of death to many of the small woodland creatures. It roams widely through the forests, slithering under and over boulders and logs and coiling up the trees. No small creature, whether bird or mammal, is safe from it. It invades the high bird nests and the hollows where the possums hide. It kills its prey by crushing it in tight coils, then opens its hinged jaws wide and lubricates the corpse with saliva before swallowing it whole.

Now the python lifts its head, tongue flickering, and casts a contemplative eye toward the little kangaroo, which gazes back transfixed with terror. But the snake decides the joey is too big to tackle and slithers off silently. The joey finds her legs and bounds back to her mother, chattering in panic.

Embedded deep within the little kangaroo is the instinctive warning to fly to her mother at every crackle of twig or small forest commotion. She will do this until she is finally able to distinguish the harmless sounds from those that signal danger. And she will acquire the necessary knowledge by mimicking her mother, sitting upright at her side, and sifting out every sound and scent with twitching ears and sniffing nostrils, because these senses can give her a much earlier warning than her sight.

Every creature in the wild must practice the same alertness or run the risk of sudden death. The calm of the bush is too frequently punctuated by the small anguished shrieks of careless animals.

A rabbit hops blithely into a thicket, squeals, and is silent; a swollen snake slithers out in its place. Rabbits are not native to Australia but are descended from domestic ones which were brought over in 1859 by early colonists for pets and food. A few escaped their pens and have since multiplied greatly until today they cover most of southern and central Australia, in some places in plague proportions.

There are thousands of rabbits in the Warrumbungles, and being rather foolish creatures, they provide food for all the carnivores. But they are also destructive, cropping away the best pasturage and moving into the homes of other burrowing creatures. In places the ground is honeycombed with their warrens.

The squeal of a rabbit at night from the dark hill slopes where the feral pigs are grunting also warns wise animals to keep their distance. The feral pigs are imports too, domestic pigs which, like the rabbits, escaped from their pens in the lowlands to make their way to the Warrumbungles as wild creatures.

Grown thinner and more muscular from their life in the forest, the feral pigs have become omnivorous. Here and there the slopes of the mountains are pockmarked with shallow hollows where they have been grubbing for roots, but they also thrive on fresh meat and carrion and have sprouted formidable tusks with which to attack enemy or prey. Even human beings have been killed by the feral pig. Though the adult kangaroo, with its swift bounds, can easily outstrip it, a young joey would have difficulty doing so. The little joey learns from her mother that the *oink oink* of a pig signifies danger and that the place where it is rooting should be avoided.

Even more deadly than the pig is the fierce feral cat. It too comes from the lowlands, where it began life as a domestic animal. It may have left its home of its own accord, or cat owners may have dumped it or its small unwanted kittens in the wild. Having to live by its wits, raising litters that learn early how to stalk and seize and kill, the cat becomes a savage and aggressive hunter. No nest of eggs, tiny burrower, or drowsy lizard is safe from its claws. If famished, it will even tackle a joey.

In its ruthlessness the feral cat has usurped the territory of the gentler breed of native marsupial cats that once stalked this wilderness, feeding on insects and mice. Unable to compete with the fearsome newcomer, the marsupial cat has become extinct in the Warrumbungles. Feral dogs, too, released as unwanted puppies into the wilderness, wreak their share of havoc.

The rangers wage a kind of running warfare against all these predators to preserve the gentler indigenous life of the wild. They trap, bait, and shoot wherever necessary, and plough a crisscross pattern through the warrens of the rabbits to trap them underground. But it is a close race and may be a losing one where some small nocturnal creatures are concerned.

A less serious threat to the balance of life in the Warrumbungles is the fox. The fox was brought to Australia in 1870, when the colonists introduced the English sport of fox hunting. The fox hunts have virtually ceased, and the fox has multiplied through the years. The lean red hunter now stalks among the eucalypti in search of anything from large insects and lizards to the smaller furred animals. His special quarry is the rabbit, but if he can outwit a kangaroo doe and snatch away her joey, he will feast indeed. The fox, however, has little chance of succeeding with the forester, which is too large to tackle, and the kangaroos usually ignore it.

A predator which is never ignored is the dingo. The forester has come to fear the dingo most because it has known this native wild dog the longest. The dingo also is an import, brought to Australia thousands of years ago by the aboriginal peoples who migrated to the continent from Asia. A domestic animal at that time, it has run wild since.

The dingo is a handsome bushy-tailed dog with a thick fur coat that is usually tawny red, though there are also cream-colored and black dingoes. The dingo mates in the fall and whelps in the winter, producing five to seven pups in every litter. It usually hunts in family packs that work together as a unit to bring down their prey.

No kangaroo tarries when the dingo's chilling bark is heard or its scent is picked up. The does and their joeys flee the farthest. The buck's size and weight cause him to tire more easily, so he can only put on short bursts of speed. His best defense is to fight and he bounds off, primarily to choose a proper battle ground, usually a stout tree against which he can back to face his foe.

Dingo against kangaroo is a close fight, and many are the stories told of the battles waged between the two. Occasionally a single dingo is more than a match for a forester. A pack is even more deadly. Sometimes the kangaroo loses his life in the struggle. Sometimes he wins. But win or lose, the dingoes must pay for their brashness. If one of them comes too close, the buck will snatch it up in his powerful arms and hold it close while ripping its stomach open with his formidable hind claws.

Sometimes the kangaroo chooses to face the dingo from an open pond. He will wade into it up to his waist and wait for the dingo to pursue him there. If the dingo is so incautious as to follow him, the buck will seize the dog and push its head under water, planting one big foot on its neck until it drowns.

Over the centuries the kangaroos have learned to be always on the alert for this predator. Even in daylight in the clearing where the mob gathers, no kangaroo feels safe. One or another of the adults is always at attention, ready to give the alarm at the first sound or scent of a dingo, though today there are very few of the dogs left in the Warrumbungles. Occasionally, however, a brief, lone howl echoes through the night, causing the doe to stiffen and the joey to shake with an ancient fright.

One day a dingo ventures into the clearing where the kangaroos are feeding, moving stealthily through the tall grass toward the doe, who is grazing near the fringe of woodland. He comes unnoticed because he does not advertise his presence; since he has chosen to approach from upwind, even his scent is carried away.

The dingo is very close before the doe finally spots him. She leaps up, chattering with fright. Clucking a warning to her joey, she is off. Head well forward, tail streaming behind her as a kind of balance, she bounds along at almost twenty-five miles an hour. Leaping frantically she leaves the joey to trail along as best she can.

Behind the doe in the clearing sounds the *thump thump* of a kangaroo buck. Alerted by her flight, he is giving the warning signal to the others by leaping up a little higher than usual and landing with a smart *slap slap* of his big feet. Then, fleeing after the fashion of kangaroo bucks, head held high instead of forward, he sails through the air, covering some thirty feet in a single bound.

*A Tragic Flight* : *111* :

Dingo Photo by Clifford Young

At his warning the other kangaroos are scattering. But it is the doe the dingo is after. Attracted by her motion it ignores the joey, who very soon loses herself in underbrush. However the dingo, which is scarcely more than a pup, is quickly distracted. When a rabbit crosses its path, it turns to easier prey.

The doe, frantic with fear, does not realize the danger has passed. When suddenly a huge heap of boulders looms in front of her, she gathers herself together and sails over it. The tension causes her pouch muscles to relax, and the tiny joey inside is jerked free. Clinging tightly to the teat, it streams along outside the pouch for a few bounds. Then its jaws are torn loose, and it tumbles dead to the ground.

Shortly afterward the doe, sensing at last that she has shaken her pursuer, comes to a halt and begins licking her arms in agitation. She listens, ears swiveling, but no sound reaches her from the forest. Slowly she begins retracing her steps to the clearing, guided by her own scent along the way.

There is an emptiness in her pouch that was not there before. But she feels more keenly the loss of her young-at-foot, and she searches for the lost joey, clucking loudly as she goes.

When a squeaky *blah blah* responds, the doe hurries in that direction. Suddenly a joey hops out of the underbrush, almost stumbling over itself in its haste. Doe and joey confront each other. Then the doe pushes the joey away. It is not her own. She resumes her slow hopping, clucking continuously.

Another squeaky voice presently answers, and a second joey bounds toward her. Again joey and doe confront each other. This time there is recognition. The doe reaches out to caress the joey's back, then doe and joey exchange affectionate nuzzles. The doe waits patiently for the joey to suckle. Presently they make their way back to the clearing.

*A Tragic Flight* : *113* :

# SIX

# *The Mating*

There is no specific mating season for the gray kangaroo, but the peak of the breeding period runs from September to March, with one young born a year. While a forester doe is carrying that young in her pouch and producing milk for it, she will not enter another breeding period. But if for any reason she loses her pouch young, she may mate almost immediately to replace it.

Toward the end of October, seven days after the little joey is torn from the doe's teat, she goes into her reproduction cycle again. Her condition is quickly recognized by one of the adult bucks in the mob, who turns his head to sniff in her direction. Quickly he hops toward her. Reaching her side he exchanges nose sniffings. The sniffings continue and include chest, pouch, and genitals.

Then the doe moves away. The buck does not force himself upon her. He continues to accompany her, quietly grazing at her side. Now and then he makes soft clucking sounds to which she does not respond; she has not yet reached the height of her cycle.

The buck is not discouraged. Occasionally he reaches out with his hands to grasp at her tail. But with a quick hop she always evades him. Outside of that movement, she continues to ignore him until she feels he is coming too close again. When she gives a raucous cough of displeasure, he backs off. He cannot force her to mate against her will. He must wait until she is ready.

On the second day the scent the doe is emitting catches the attention of the two other adult bucks. They too start bounding over. The first buck sees them coming. Quickly he rises to his feet, determined to fight for his mating rights. Standing on tiptoe and balanced on his tail, he crosses his arms in front of him—an aggressive action. His head turns from side to side as he licks and bites at the fur on his upper chest, all the while giving the same soft clucking sound with which he has been pursuing the doe. This time it is a challenge to his rivals.

The nearest buck responds by rising to his toes also. He coughs sharply, accepting the challenge, and the fight begins. The doe stops grazing and sits up to watch with interest, her joey perched beside her. Other does and their joeys are watching too.

This is no friendly contest. The powerful bucks clasp each other around the shoulders and wrestle together, trying to topple each other over. Neither succeeds because they are so evenly matched.

Now balanced on his tail, each tries to hug his opponent tightly to his chest, at the same time using his powerful feet to aim accurate kicks at the lower body. The deadly claws on those feet could easily disembowel an adversary if directed downward. But the bucks deliver their blows in straightforward thrusts when they are fighting other kangaroos. Their chief aim is to overthrow their opponent, not to kill him.

Meanwhile the third buck, who tries to take advantage of the situation, is suddenly confronted by a fourth male that comes bounding out of the forest to join battle. As the two lock together in struggle the doe gets bored and drops down to graze a little. Then she lets her joey suckle, after which she returns to her ringside seat.

The struggle between the first two bucks is growing fiercer, the rasping coughs sound more sharply, and the blows fall with more

Koala - Photo by Barry Muir

Wallaby

Pigmy Possum

Marsupial Mouse

deadly force. The claws of the second buck rip through an ear of the first buck. Furiously he responds by lashing out with his clawed foot, peeling a strip of fur from the belly of his opponent.

Scratch! Claw! Kick! Tear! But never use the teeth! Presently blood is streaming from a dozen lesser wounds in both animals. Sometimes a duel for mating rights may end in death for one of the contestants, but this one does not go that far. The second buck finally just drops to his haunches and bounds away. The first buck watches him go, twitching his ears.

The other contest has already come to an end almost before it started. The newcomer quickly gave up and disappeared into the woodlands leaving his opponent free to court the doe. He approaches her and sniffs. She returns the sniffing, but when he reaches for her tail, she hops away. She will mate with none of them till she is ready.

The first buck starts grooming his matted, bloody fur. Then he looks around for the doe. There she is, pursued by a rival. In a fury he bounds over to issue his challenge, and another fierce struggle begins.

This fight does not last long. Quickly moving his muscular forearms, the first buck topples his opponent to the ground. But he is too angry to stop there. He continues kicking at the fallen kangaroo until it drags itself off with slow painful hops.

The triumphant buck once more turns his attention to the doe. As time passes he becomes more and more ardent in his pursuit. He reaches out for her tail with quicker, more aggressive movements which she is no longer so eager to evade. By the third night she is ready to mate. Instead of hopping away from the buck, she stops and crouches forward in front of him, her hands on the ground.

The buck moves up, plants his feet on either side of her tail, and grasps her around the chest, pulling her to himself. As he mates with her, his arms shift to her abdomen, where he clasps her in a close embrace.

During the next hour or so the buck mates with the doe a number of times. Every now and then another buck will come bounding out of the forest, attracted by the scent of the doe, and will try to push the mating kangaroo away from her. Then everything has to come to an abrupt halt so that the intruder can be driven off.

Finally the doe pulls away from the buck. The mating is over. But the buck and doe stay together throughout the rest of the night, grazing or resting side by side with the joey nearby. Occasionally the buck sniffs the doe, clucking softly. She returns his affection by gently patting his ears.

But with the break of day, the buck loses all interest in his mate of the previous night. One whiff floating down the air from another doe whose joey has just vacated her pouch for good tells him that she too is nearing her time. Off he bounds, just as eager to mate with her, even though he may have to fight fiercely again for the right to do so.

The doe he has left behind is indifferent to his going. The peak of her breeding cycle has passed and new life is planted within her. In little more than a month there will be another embryo to take the place of her lost pouch young. For the present she will pursue her daily activities as usual.

# SEVEN

# March of the Seasons

As the season warms, the doe and joey gradually shed their warm winter coats for thinner ones. When the mid-November summer slams down hot and brassy, they begin coming to the clearing long before dawn and retiring more quickly to the shade. But even the shade is thin and sparse because the leaves of the eucalypti preserve moisture by hanging down instead of spreading horizontally, turning their thin edges sideways to the sun.

On hot humid days when the air clings close and heavy to the earth, the kangaroos tend to keep apart from one another, and they become irritable as the ticks and insects grow more active with the warmer weather. Swarms of humming mosquitoes dive-bomb their faces and ears. Clouds of sand flies cluster on their eyelids, at times biting so severely that a few foresters are blinded and are left helpless to die.

To drive off the insects the doe and her joey scoop out shallow resting places in which to lie. Picking up handfuls of dust, they toss it over their bodies. Peewees, willy wagtails, and flycatchers also come to their aid. Perching on the backs and rumps of the kangaroos, they launch themselves at intervals upon the swirling insects, which they devour daily by the thousands. But the birds exact a price for their services, yanking out beakfuls of fur to serve as lining for their nests.

The voice of the summer is the voice of the cicada. Throngs of the noisy insects rend the air with a multitudinous shrilling from high in the treetops where they are feeding. The piercing, pulsating chorus begins with the early morning light. Broken by brief pauses, it increases as the day wears on until, by noon, it is almost deafening.

With every passing week more and more nests are being deserted as birds reach the fledgling stage and take to their wings. A mother quail hurriedly shepherds her young out of the way of the hopping kangaroos. A father emu drums his brood of half-grown chicks across an open clearing.

By mid-December the sky over the Warrumbungles is full of the young of the wedgetailed eagle. Their light brown juvenile coats contrast sharply with the black plumage of the great adults. Cousin to the golden eagle of the United States, the wedgetails are huge birds three feet long with six-foot wingspans. Rolling and tumbling through the sky, riding the wind currents created by the high peaks, suddenly dropping a thousand feet in a minute, family groups engage in aerial sports together. But soon the adults will be harrying their young from the Warrumbungles. The juveniles will spend two to three years wandering before they settle down in territories of their own.

Toward the middle of December the doe and her joey, along with all the other foresters in the Warrumbungles, begin an upward trek from the valleys and lower slopes. Vacation time has come to the downunder land, bringing with it thousands of humans, and the kangaroos are retiring from the influx.

The doe and her joey travel upward in leisurely bounds. On the hot northern slopes of the mountains, cypress pine, ironbark, and bloodwoods rise above almost impenetrable daisy thickets whose dark green leaves are coated with a sticky varnish. But on the cool southern sides large white box, silver tops, and stringybarks shade the grass clearings. Water can be found in rocky pools and in the Wombelong Creek, which after rainfall dashes over its boulder-strewn bed on its way to the valley.

Here the kangaroo mob begins to gather again. All about them the remnants of ancient lava flows rise in peaks, bluffs, cliffs, and rocky spires. Seldom does a forester penetrate these heights, which command a view of the vast plains that surround the Warrumbungles like a sea flowing away to a far horizon and the Indian Ocean 2,000 miles distant.

Over these plains roam two cousins of the forester—the western gray kangaroo and the red kangaroo. But they never meet, for the forester does not leave the Warrumbungles while the red kangaroo and the western gray are bound to their semidesert homelands.

Gale winds, scorching in summer, freezing in winter, blow across the plains to buffet the exposed heathlands which in places cover the heights. In this poor, thin soil only the hardiest vegetation can take root. There are dense thickets of mallee, a form of stunted eucalyptus with a great rootstock from which rise a number of stems instead of a single trunk. Fringed heath myrtle and bushy whitebeards with their tiny bellshaped flowers are interspersed with open stretches of snow grass.

These wild craggy lands are the favorite haunt of the gentle rock wallaby. Smaller than the wallaroo, the rock wallaby also has padded feet fringed with coarse hair. Its tufted brush tail acts as a balance when it leaps from ledge to ledge, agile as a monkey.

The rock wallaby feeds mainly on the high heath; in the heat of the day, it retires to the deep cool mountain caves. Its ancestors have used these caves for so many thousands of years that the stone floors are worn to a high gloss by the movements of wallaby feet. Many of the caves, however, have now been lost to the far

*March of the Seasons* : *121* :

more aggressive goats. Introduced to Australia at about the same time as the fox, they have since run wild, penetrating even to the rocky slopes of the Warrumbungles. Here they race fleet-footed along the faces of the steepest cliffs or lounge in the ancestral caves of the wallabies. On the hottest days the wallabies, driven from their shelter, often succumb to heat stroke.

As summer draws to a close, the doe gives birth to her second embryo, sitting patiently while it makes its way to her pouch to begin its slow growth to maturity. In two weeks' time it reveals its male sex.

The signs are now all pointing to fall. The droves of summer vacationers disappear. Fall rains enable the farmer who owns the high fields in the Warrumbungle foothills to plow and sow oats on which he will graze his livestock.

Fall does not bring a great migration of birds, though the bee eaters return to their winter homes, and honey eaters like the lorikeets follow the blossoming flowers toward the tropics. But the crimson rosellas will stay the year round, finding enough seeds to sustain them through the winter, while the budgerigars will spend their days on the plain below, feeding on the ripened seeds of the wild grass. And the galahs, much to the farmers' annoyance, will follow the fall plantings to snap up any spilled wheat.

Wherever they may be spending their day, the birds always return to the Warrumbungles in the late afternoon, flying in by the thousands to drink at the sweet mountain pools. First to arrive are the budgerigars. They come in bands of thirty to forty, hitting the bank of a pool in unison as if at a signal. Each scoops up a beakful of muddy water. Then they are off again, circling like flashing emeralds in the sun to evade the brown hawk swooping low.

Next come the galahs in an explosive cloud of gray and pink feathers, squabbling and squawking among themselves. Last of all

Just hatched Emu chicks

On a warm day Kangaroos seek the shade

Young emus follow their father foraging

the stately cockatoos fly in. The great white birds perch in a row on the bank making noisy conversation while they drink. Then they, too, are off.

As the season progresses, a chill tang sharpens the air. The trunks of the dwarf gums on the heights herald the approach of winter by turning a vivid red. Sometimes in the early morning a thin rind of frost coats the grass.

Everywhere the wilderness creatures are preparing for winter. Many lizards stop eating and seek refuge under the loose bark of the eucalypti until warm weather comes again. But the anteaters and the snakes will go into complete hibernation, as will the shingle-backed skink. Already its tail is bloated with stored fat upon which it will draw throughout the winter.

The little fat-tailed marsupial mouse is also using its tail as a storehouse, though it will hibernate only for four or five days at a time during the coldest spells. Its tail is now so puffy with fat that it looks like a large carrot.

The ringtailed possums are starting to prepare new nests for the litters which will be born in winter. The possums simply add another nest to the collection of large, domeshaped ones they already have. In these collective nests or dreys, as they are called, several generations live together. Gathering leaves, ferns, and twigs, the possum lays these building materials on its tail, then coils the tail around the bundle and is off to its worksite.

Both possums and kangaroos are exchanging their thin summer coats for thicker winter ones. In the pouch of the forester doe, the embryo is developing daily, creating a growing bulge which can now be plainly seen. Its body is still practically naked and its eyes will not open for almost another month. But sometimes it can be heard squeaking inside the pouch.

Far more telling changes are taking place among the young-at-foot in the mob. All through the summer their periods of suckling have been stretching farther and farther apart. Throughout April and May they are being weaned altogether. Now they are fully able to care for themselves, though they still stay close to their mothers.

As joeys they all looked a great deal alike, but as they leave babyhood behind, they begin to develop rapidly. The young male joeys start putting on weight; they rapidly become taller and more muscular. Frequently they spar with one another to test their strength.

The female young-at-foot grows much more slowly, and her once-playful antics become subdued. At eighteen months she is no longer a carefree joey but a sedate young doe old enough to mate and have offspring of her own. Her wanderings through the wild at the side of her mother have taught her how to fend for herself. She has learned what creatures are harmless, what dangers she must flee from, where the choicest grazing grounds and water holes are. Now and then she will leave her mother's side to make short exploratory excursions on her own. But she will always return.

The bond between mother and daughter has grown too strong to be broken easily. It is the longest lasting association that kangaroos experience.

# EIGHT

## In the Farmer's Field

A bleak world greets the younger joey when, at six months of age, he first pokes his furry head above the rim of his mother's pouch. It is July, the dead of winter in the Warrumbungles. The high peaks are dusted with snow which has lingered for a week on the southern slopes where the sun never shines. The temperature is a freezing eight degrees Fahrenheit, a normal reading in the Warrumbungles, which can be the coldest place in Australia in the winter time.

The little joey greets the freezing air with a perplexed frown. His ears, unlike those of the adults, flop over his forehead. His nose shivers with cold. In an instant he pulls back his head and sinks down again into the luxurious warmth of the pouch. Grasping the teat, he suckles greedily.

The air is crisp and dry. There are days of cold sparkling sunshine; at sunset, the sky grows a deep rich blue. On the horizon the great luminous globe of the sun sinks slowly away, casting as it goes a rich, golden glow upon the high rocky spires.

The sky darkens. Stars glitter or a great moon shines down whitely on glistening meadows rimmed with frost. On the south side of the peaks, the Castlereagh River and the Wombelong Creek both wear thin crusts of ice.

During the warmer days the frost turns to sludge. Then in the bitter nights it quickly freezes again and more frost forms. As the weeks go by, a two- to three-inch shield of ice builds up, coating the ground.

The yellow-footed marsupial mouse chooses this season for his fierce matings. Males quarrel frantically among themselves as they vie for the females. Even the mating itself looks like a battle. The male grabs the smaller female by the nape of the neck, sometimes inflicting serious wounds on her head and neck as he twists her into position. The mating lasts from five to twelve hours, but the male pays for his prodigious expenditure of energy. By the time the young are born one month later, their fathers will all be dead. The females will live a year or two longer.

The fierce wedgetailed eagles are also engaging in their yearly courtship rituals. High in the sky with claws interlocked, pairs of the great birds tumble through the air performing astonishing acrobatic stunts that end in mating.

During the cold winter months the kangaroos choose sunny rather than shady spots in which to lie, and they draw closer together for warmth. There is little that is appetizing in the frost-burnt grass on the clearings. But down in the high foothill fields the oats are sprouting fresh and green.

Drawn as if by magnet, the hungry kangaroos make their way to the foothills and those lush fields fringing the reserve. Within its boundaries they are completely protected. They also have some degree of protection outside, especially in the state of New South Wales, which allows no one to shoot a kangaroo, even on private property without first obtaining a permit. This is granted only if an investigation shows that the animals are causing damage.

Though the kangaroos cannot know that their feeding in the fields will qualify the farmer for a permit they seem to sense that there is danger here where the pasturage is so fine. The odor associated with that danger belongs to man.

As they reach the open fringe where woodland and fields

merge, they become very wary. Keeping to the edge of the field, they nibble only briefly at the young sweet plants, pausing often to sit upright, listen, and sniff the air. If any of them should catch one whiff of human scent, the boomers would thump out a warning and the entire mob would bound away into the underbrush.

But nothing happens, and with the passage of time the animals become bolder. For two days wallaroos, kangaroos, and red-necked wallabies browse along the edges of the field unmolested.

Then on the afternoon of the third day, a farmer stealthily approaches, gun in hand. Not a single kangaroo recognizes any threat in him because he has taken care to come from upwind, and his scent has been carried away.

The farmer has obtained his shooting permit, though the rangers have tried to dissuade him from using it. They point out that kangaroos, unlike cattle and sheep, only graze off the tops of the oats without injuring the plant itself. Kangaroos also eat only what their system requires, a far smaller quantity than cattle and sheep. But the arguments have proved fruitless. The problem lies not in what they eat, the farmer says, but in the havoc they cause in his fields with their big feet and tails.

Now he lifts his gun and fires. The clatter ricochets off the hills, stampeding the kangaroos. A great buck falls, kicking feebly. Another buck leaps high in the air and lands with a resounding thump. Before he can make another move the gun roars again, and he too collapses.

The doe and her daughter take to flight. Behind them a wallaroo doe is hit in the stomach, and the joey in her pouch squeaks with pain. It too has been struck.

A bullet nicks the ear of the doe herself. Then she is in the screen of woodland and thicket, her daughter close beside her. Once again within the parklands, they are safe. But as the shots continue to rattle behind them, they keep bounding along until they reach a familiar clearing. Here, shivering with fear, hearts thumping, mother and daughter stop to lick their arms distractedly. Blood trickles from the ear of the doe, and she lifts a hand to brush it away.

Suddenly a big buck leaps into the clearing, startling them to flight again. But the buck stops short. His chest heaves as he struggles for breath. All at once a shudder runs through him. He topples over and is still, though there is no mark of any kind upon him.

For all its size and strength, the kangaroo has a nervous system so finely strung that it cannot endure prolonged stress. Being heavier than the does, the bucks become more quickly winded and react more acutely. The strain has been too much for this buck, and he has suffered a fatal heart attack.

Within a few minutes the meat ants are swarming over his still body. A fox steals out of the underbrush for his share of the carrion. Crows and ravens flap down for a feast and are joined by a kite, though these predator hawks are not usual companions to crows. Even the magnificent wedgetailed eagle condescends to take his share of the bounty. In a week or so, little but a few bleaching bones will be left to mark the end of the great buck.

# NINE

## *The Wedgetail's Attack*

The little joey in the pouch has been growing bolder with every passing day. Now his head appears above the rim of the pouch more often and stays out longer. While his mother is grazing, he finds that by sticking his own head out he can reach the grasses below him. He begins to lip them, drawing the blades through his mouth from one side to the other in mimicry of his mother. One day he will translate this play into actual chewing.

By late August the joey is eight months old and weighs about eight pounds. He now measures more than a foot and a half from the tip of his nose to the tip of his tail. He has become so large that the doe's elastic pouch hangs like a bag of potatoes in front of her abdomen. He is much too heavy to be carried about constantly. Besides, it is time for him to start practicing life outside the pouch. One day the doe gently tips him from it with her forehands.

The joey's weak legs cannot hold him upright, and he sprawls out flat. Desperately he tries to get to his feet, pawing at his mother's abdomen to find the pouch. But she sits very straight so that the rim is above his reach. After a short while, his mother decides he has had enough for the time being and is ready to let him back in again.

Giving her soft clucking sound, she bends forward so that her pouch is closer to the ground. When he still cannot reach it, she pushes her hind legs backward to lower her pouch even further. The joey begins groping along her neck and chest as he searches for the rim. When he finds it, he holds it open with both hands and tumbles in headfirst, long legs and feet thrashing in the air. There's a shaking and quivering in the pouch, the legs disappear, and out pokes the head again. The joey has made a complete turnabout.

Several times a day the joey spends short sessions outside the pouch. Presently his muscles begin to gain strength. One day he takes a hop or two, then several more. But he never strays far from his mother. A sudden noise of any kind, a burst of wings overhead, or the guffaw of a laughing kookaburra sends him rushing back for the pouch.

But the doe will not humor him. She insists he get acquainted with the outside world. If there's no real danger, she won't let him in, sitting upright so that he can't reach the pouch. All he can do is stand on his hind legs and thrust his head inside. Even this is comforting to him though, and he stays there until the darkness bores him, then out pops his head again.

During the chilly night hours when the doe and her daughter graze together through the forest, the joey rides secure in his mother's pouch. Even after they have returned to the clearing, she will not let him out until the sun has dried the grass.

When the dew disappears, the doe stops grazing and begins to clean her younger joey, her soft tongue licking over his furry head and ears. Then she tips him out to finish the job. But the joey has had enough of grooming. Quickly he turns it into a wrestling match. Jumping to his toes, he grapples his mother around the neck and reaches for her ears. She has to lift her head high to protect herself from his small sharp claws.

Unsuccessful at grabbing her ears, the joey next tries to give them a kick with his flying feet. All he manages to do is topple over in a heap. Then he's off to amuse himself with a little game of leaf-waving or twig-throwing. He picks up a twig and tosses it as far as he can. Then he hops after it and tosses it again, over and over, until he tires of it. He goes back to jumping around and around his mother in short bouncing hops like a rubber ball. When she starts grazing, he tries to snatch away the bunch of grass which she's chewing.

Other joeys like himself are hopping about the clearing in all directions, popping in and out of their mothers' pouches. When a sudden threat of danger sends the kangaroos scattering, each doe collects a joey, perhaps her own, perhaps another's, and is off. It is impossible for an outsider to tell if the joeys have been sorted out properly, for they all look alike. But every mother knows her own offspring.

By this time the male joey has grown so big he no longer needs to climb into the pouch for his milk. He just stands up on his hind legs; with his hands on either side of the pouch rim, he holds it open and buries his nose in the dark interior to get at the teat, which has been growing longer with every passing month.

Sometimes the joey mistakes his nearby sister for his mother and tries to reach into her pouch, too, for a guzzle of milk, but she cuffs him away gently.

This is a time of great peril for all the wilderness young because as winter moves into spring the wedgetailed eagle's white eggs, two in a clutch, begin to hatch. Now the ravenous eaglets are clamoring for food. They are so rapacious that almost half of them will be slaughtered and devoured by their own nestmates before they become fledglings. The parents are kept hunting from early dawn to dusk. Everything is fair game for the desperate birds— rabbits, possums, lizards, kids, and joeys.

The joey is unaware of the hazard. The big birds in the sky

This teenager, who is a member of a volunteer group called Marsupial Mothers, is helping to raise a kangaroo joey whose mother was shot while it was still in the pouch.

mean nothing to him. Every day he becomes more independent. He rebels against staying in the pouch for long. The space is becoming more and more cramped. Sometimes he finds the most comfortable way to sit is on his rump with hind legs, forearms, and head all protruding.

One early morning as his mother and sister are making for the familiar clearing, he simply tires of it all and tumbles out of the pouch. Off he goes bouncing into the forest. In a patch of open ground, he begins to play a game of tag with his own shadow.

Suddenly his small shadow is swallowed by a large, dark one. Looking up, he finds himself staring into the golden eyes of a wedgetailed eagle. Talons extended, hooked beak pointed like a rapier, it is swooping down upon him.

With a bleat of terror the joey bounds away. He looks around for his mother, but he cannot see her anywhere. He is alone. Terrified, he hops in a zigzag course across the small clearing. The diving eagle swoops down, narrowly missing him. The wind in the great pinions whistles through his ears.

The momentum carries the eagle along for a while. Then it circles back for another dive. The joey, now utterly confused, begins hopping round and round in circles, bleating instead of making for shelter.

The doe answers his cry with loud clucking sounds. But the joey does not know now in what direction to run, and he is tired, so very tired. More and more slowly he continues his aimless hopping.

By the time the doe reaches the clearing, the eagle is diving again. With a bound the doe plants herself between the joey and the plummeting eagle. Vicious talons rake across her slender shoulders, drawing blood. Then the eagle is lifting again.

Once more it swoops down. This time it tries to drive a wedge between the joey and the doe. But the joey has finally found his bearings. He stops circling and makes for the underbrush. Behind him the doe flails out at the eagle, kicking and clawing at its breast and wings.

The respite gives the joey a chance to reach the shelter of the thicket. And the frustrated eagle soars away. The doe begins clucking to the joey to join her, but he is too terrified to leave the sheltering shrubs. Instead he crouches there, uttering a rising chorus of bleats.

His plaintive calls are too much for his sister. She comes bounding through the shrubbery in search of him, for it is instinctive in kangaroo does to rush to the aid of crying joeys.

When the joey finally scrambles out of the brush, he is too bewildered to recognize his mother. He hops to and fro from her to his sister, looking for the pouch. Finally he recognizes his mother's clucking calls and bounds to her. Into the pouch he tumbles head-first, his long legs waving. Soon the legs are still, sticking stiffly out of the pouch like crooked sticks. The joey is too worn out from his ordeal even to turn himself right side up.

# TEN

# *The Great Fire*

Ever since early winter the weather has been exceptionally dry. Usually some thirty to forty inches of rain fall every year on the Warrumbungles, mainly during the summer and winter. In the fall and spring, dry periods of a month or so are common. But occasionally a longer drought closes like a vice upon the mountains. This has been such a year.

So when spring arrives with a refreshing rainfall, the kangaroos hurry to mate, because rain means rich pasturage which will provide the nutrients necessary for the nursing mothers. Almost all the does enter their breeding period simultaneously.

The young males who were weaned the previous fall and are still tagging along with their mothers suddenly find themselves outcasts. The mature bucks that come courting are jealous of the youngsters and drive them away. From then on, the juveniles will either live alone or band with others their age, forming units of two or three. If they choose they may join another mob. As long as they do not appear to be asserting mating rights, they will not be molested.

There are other bands made up of elderly foresters, which also avoid the society of the mobs. These old bucks are all past twelve years of age. There are not many such bands because though foresters have a potential life expectancy of twenty-three years, half of them die before they reach three years of age, and only fourteen percent will ever pass seven years. Drought, disease, and animal and human predators all take their toll.

The kangaroos that do survive these hazards face yet another —advancing age. Time deals harshly with the oldsters. The gloss fades from their coats. Their teeth fall out one by one, making grazing a slow, time-consuming process. Some retain only one or two worn rear molars in each jaw.

The old bucks have also lost their powers of reproduction and their urge to mate. So, shunning the excitement of the mobs, they are content to browse and loaf their days away among their own company.

There is no such respite for the doe whose life expectancy is as long as the buck's. She will mate and bear offspring to the end of her life, though after sixteen her fertility will diminish. Both the doe and her daughter mate this spring, and immediately new life begins to develop in the daughter's uterus.

It is different with her mother because the young joey still uses her pouch and normally would do so for another two months. Usually the doe waits until the joey has permanently left the pouch before she mates. But because of the year's unusually arid conditions followed by rainfall, she has mated early. Now she must finish training her joey to live entirely outside the pouch before she can give birth to her embryo, or she will endanger the new life.

Her system automatically takes care of the situation by stopping the growth of the embryo after it has developed to a tiny bundle of some eighty-five cells. This group of cells, which is called a blastocyst, will remain dormant until the pouch young leaves, after which it will complete its development. The forester does rarely have dormant blastocysts. The western grays apparently never do. It is only the red kangaroo that habitually produces a delayed blastocyst. This follows her mating shortly after the birth of an embryo. She then carries the blastocyst in readiness for the time when the joey in her pouch leaves it for good, after which the blastocyst resumes its development.

Now the forester joey's training begins in earnest. He no longer finds such a complacent mother. Often when he wants to return for a snug rest in the pouch, she holds him off with forearms crooked in front of his haunches.

Sometimes he breaks through her defenses and dashes for the pouch, managing to get his head and shoulders inside it. But before he can go any further, the doe bounds away, dragging him along unceremoniously until he drops off. If he does manage to enter the pouch, he no longer finds it the comfortable haven it once was. The doe starts hopping around vigorously, purposely bumping him against boulders and tree stumps, or she simply relaxes the elastic rim and lets him take a nasty tumble. When nothing else works, she cuffs him.

But there are other times when the doe relents and welcomes him back to feed and rest. These times, however, are becoming more and more infrequent. By the time the joey is nine months old, he has become accustomed to living outside permanently, though he continues to suckle from the teat which is reserved for him. He has become a young-at-foot a month and a half earlier than did his sister.

With the joey gone, the stretched pouch rim which had developed weeping sores during his occupancy quickly heals, and the whole pouch regains its former elasticity. All this is in preparation for the new birth, which takes place a little more than a month later, as another embryo makes its way successfully to the doe's pouch.

The joey is entering a desolate world. After that one brief rain the weather becomes dry again. The wild flowers that sprang out quickly wilt. And honey eaters are wandering far afield for their nectar.

The grass in the clearing turns sere early in the season, and the kangaroos separate to look for better pasturage. No use now braving the farmer's gun to enter the foothill paddocks. They, too, are dry and brown, and the cattle grazing there look gaunt.

Water is also becoming harder and harder to find. The small streams have dried up, leaving behind only beds of bleached pebbles. The few puddles which still remain in the larger waterways are mere shallow seepages.

But in the Warrumbungles water can always be found somewhere. Even in the most barren stretches the doe and her daughter and joey come upon pools noisy with birds. These pools are fed by springs trickling out of clefts where the volcanic rock meets sandstone. There are many such springs in the mountains, and they flow from sources underground so deep that they never run dry, even in a great drought.

The pool is something to quarrel over. Sometimes the does and the joey come to one of the small rocky basins of green water only to find a big buck trying to curl himself into it. He lies first on one side and then on the other, splashing his face, chest, and shoulders with flailing hands. When the does and the joey try to get a drink, they are driven off with a cuff or a kick.

Day follows blistering day. Even in the early morning the sky hangs over the earth flat and hard like a rust-colored porcelain plate. The sun brings temperatures that soar well over one hundred degrees Fahrenheit. The cypresses are dying. But the phalanxes of eucalypti with their heat resistant systems seem unaffected, though their leaves droop heavy with gray dust. Clearings and once-grassy slopes are now maned with inedible yellowed grass, and the twigs and branches of the brittle underbrush rattle together like strands of wire.

With most of their pasturage gone, many of the rabbits begin to die. At first they create a glut for the forest's scavengers. Then the carrion is gone, and starvation again resumes its slow course. Only the eagle can escape. It roams farther and farther afield in search of food, making sweeps over the sheep lands in search of ailing lambs too weak to fight it off.

Smaller birds cannot forage so far, and their fledglings die in the nests. The eggs of the lizards are drying up. The once-puffy shinglebacked skinks are now so thin they can scarcely be recognized. Many will not live to see the autumn.

The coats of the kangaroos look lusterless. Their bodies are gaunt, their ribs protruding even more than usual. The bone at the base of the tail is plainly evident, a sure sign of malnutrition. The forester, used to more congenial weather conditions than the red kangaroo of the plains, cannot endure drought conditions. Eight months of poor pasturage will dry up the mothers' milk supplies and take a heavy toll of the young.

As the hot, dry weeks slip by, the little embryo in the doe's pouch dies, and she tips out the shriveled, grublike creature. A month later her daughter loses her young. All through the mountains other does are having the same experience. None will reenter the mating cycle until well after the drought is broken, when the ground is green again and they have regained their strength.

In the midst of all this aridity, some species of eucalyptus suddenly burst into flower. The rangers who have seen this phenomenon repeated over the years know that the out-of-season blossoming is a sure herald of coming rains. But when will they arrive?

At the end of summer, heavy clouds from the north advance to blacken the heavens above the Warrumbungles. The threat of lightning hangs over the forests. In Australia's national parks there are hundreds of lightning strikes every year. Each one can start a fire; most of the fires do not spread because they are put out almost immediately by the accompanying rains. However, once every forty years or so, a great conflagration occurs. This usually happens when, during a drought, the forests are subjected to dry storms.

One day a series of such storms comes sweeping over the Warrumbungles. No rain falls. But suddenly out of the black clouds lightning streaks down time after time, its crooked rapiers plunging into the forest. Thunder cracks sharply. With every peal a quiver runs through the nervous mob.

Here and there as the lightning finds its mark an oil-impregnated eucalyptus bursts into flame and is instantly enveloped in smoke. Presently a number of coils of smoke appear over the mountain slopes. Under the smoke the fire rapidly spreads and intensifies. All at once sheets of orange flames break into the open, raging through impenetrable thickets and leaping from tree to tree.

Winds created by the heat bear fiery scrolls of eucalyptus bark through the air and drop them into the bush. New blazes spring to life. The volatile oils in the burning eucalypti gather in drifting gaseous clouds until they become so overheated they explode, spreading the fire. Their coats aflame, rabbits race out of grassy clearings and ignite fresh brush before they die.

By nightfall the flames have reached the ridges. Tongues of fire race along the humpbacked mountains with a crackle that can be heard a mile away. Millions of glowing sparks flitter everywhere while the downcurrents of wind carry the choking smoke into the valleys.

The only escape for the small creatures of the wild is to force themselves under rocks or into deep burrows, but few succeed in doing this. Larger life is on the run. Kangaroos, pigs, goats, dingoes, and foxes all crash in panic through the undergrowth. Dense smoke envelops the doe as she bounds along, clucking to the joey to follow. For a while she hears his *blah blah blah* behind her. Then suddenly his crying is lost in the holocaust. He is nowhere to be seen. Her daughter too has become separated from her. But the doe, confused and frantic, does not miss them immediately.

*The Great Fire*  : *141* :

On she bounds, eyes streaming with tears from the thick smoke. She draws scorching air into her bursting lungs. The crackle and snap of burning trees is all around her as she weaves a crazy way among their trunks. As one or another suddenly flares up before her, she has to change course from one bound to the next.

Glowing embers fall on her fur with a quick sizzle, leaving the stench of burning hair and flesh. The ground underfoot has become so hot it scorches the soles of her feet. Drifts of soft, hot ash impede her way. Occasionally she hears the *thud thud thud* of other kangaroos as terrified as herself. But she sees none of them.

The fire is not burning unopposed. Weary, soot-blackened volunteers work with trained firefighters and rangers to create firebreaks in the hope of containing the flames. But despite their efforts, the fire burns on for four days.

At last the heavy front of rain predicted by the flowering eucalypti comes down from the north. It pours in torrents upon the tortured mountains and finally drowns out the fire.

In an untouched gully far from her usual haunts, the doe stops her panicky flight. The forest around her is unfamiliar. As far as she can see, there are no other kangaroos nearby. Loneliness envelops her. For a while she hops around, distractedly clucking for her joey and her daughter, to whom she has been so close for so many months.

There is no answer. The joey is gone forever. There is no way of knowing whether the daughter has survived; at least she is nowhere near. At last the doe's clucking fades away. She stops her aimless hopping. As if aware for the first time that water is nearby, she stoops and quenches her parched mouth and throat in a puddle.

The quiet emptiness of this foreign woodland oppresses her, and the sense of loneliness and loss increases. Bedraggled, disconsolate, wet to her painfully scorched skin, she stands quivering in the driving rain—the rain that came too late.

Fleeing Kangaroos

In the wake of the great fire

# ELEVEN

# *Renewal*

With the passing days the doe wanders alone through the strange forests, browsing and resting, exchanging sniffings now and then with kangaroos she meets. They are all unfamiliar to her. None even come from her section of the forest.

Many of the kangaroos were, like the doe, able to outrun the flames and escape. Others were trapped or lost their way in the smoke and circled back into the fire. Some were asphyxiated. Some died of heart attacks. Many were stampeded onto private lands where they were either tolerated by sympathetic landholders or shot by others fearing the wholesale destruction of their property. A number broke their necks or legs on the high wire fences which enclose some farms and grazing grounds.

The kangaroos that have survived are exiles like the doe, for their once-lovely home is a desolation. More than twenty-five thousand acres of wilderness have been burned away, exposing the bare bones of the land. Gaunt, blackened trees rustle their amber clouds of bleached leaves. The charred remnants of other giants sprawl across the earth.

Fire has continued what the drought began. Many tiny burrowers are entombed in the homes where they sought shelter. The charred bodies of sugar gliders curl round one another in the hollows of blackened trees. Roasted koalas still crouch in the high forks of the eucalypti where they were caught unaware. In dense, flame-blasted thickets the corpses of ringtail possums still swing by their tails, and the shriveled ribbons of pythons coil in loops from the trees.

The carcasses of larger animals are strewn where they fell in midflight. An oppressive silence has laid a heavy hand upon the whole land. No bird or insect breaks the stillness. There is only the sound of water rushing in torrents over the naked earth, leaving its mark in new gullies and eroded slopes.

But in the days that follow, rains intermingle with sparkling sunlight. And within weeks, life begins to reassert itself. Fresh new grass springs up everywhere. Under stands of burnt-out cypress pine and acacia, heathland plants sprout.

The eucalypti do not die so easily as other trees. Their scorched trunks and branches soon put out bunches of green leaves called epicormic shoots. These will enable the trees to survive until they can clothe themselves anew in proper foliage.

The fire has brought life as well as death to the forest. The hard casings of acacia seeds which have lain in the ground for years, waiting for the holocaust of fire to crack them open, now split apart, giving the kernels within a chance to swell and grow. Everywhere acacia saplings are springing up intermingled with young cypress. As time passes the epicormic growths of the eucalypti give place to fresh foliage. The underbrush renews itself.

The wakening plant life first draws swarms of insects. Then come flocks of insect-eating birds. Rabbits return, drawing in their wake the eagles and the hawks, the dingoes, the foxes, and other

The doe comes home to a renewed forest

feral carnivores. Lizards, burrowing creatures, and small marsupials gradually repopulate the empty wilderness. And kangaroos as well return to feed on the fresh green meadows.

One day the doe feels a restlessness growing within her. She turns back toward her old haunts and comes again to a familiar clearing. Other foresters are already there. She has never seen any of them before, but she greets them with a round of sniffing and then falls to grazing. Easily she enters into the old rhythms of her life.

Yet there is a difference. She has no young-at-foot, and there are very few other joeys hopping about lightheartedly, making nuisances of themselves. Most of the youngest generation of kangaroos has been lost either in the fire or the drought.

But this will change presently. The pouches of many of the does are already swelling with growing embryos. Soon the doe herself will mate again and bring forth another life to face the many hazards of a kangaroo's existence.

On this she does not ponder. The grass is sweet to her taste. High in the reclothed branches of the fire-blackened trees where the downunder September spring has brought out fresh blooms, the noisy friars are squawking. In the distance a kookaburra is laughing. Everywhere life goes on.

# *Epilogue*

Of all the continents in the world, Australia has the richest variety of marsupials. They range in size from the tiny pigmy glider and the yellow-footed marsupial mouse to the large gray and red kangaroos. Many of the small marsupials are burrowers, and since they all come out primarily at night, few humans have seen them.

The largest marsupial predator in history is the thylacine or Tasmanian tiger, which hasn't been observed for years, having been hunted to the verge of extinction by early farmers. Hopefully a few of the strange creatures, which look like striped dogs, still roam the wilds of Tasmania.

On the other hand, the Tasmanian devil still flourishes. Though it appears quite ferocious with its huge head, powerful shoulders, and jaws, it is in reality a rather clumsy and harmless scavenger. It can devour a whole carcass, including the bones, leaving behind only the teeth.

There are four species of native marsupial cats and one marsupial mole in Australia. Some ten species of bandicoots, also marsupials, make their homes in and on the ground. They get their name from a species of large Indian rat which they resemble. The bandicoots are quaint-looking creatures. One species has large, rabbit-shaped ears and another a very long nose. A third, probably extinct now, has feet shaped like a pig's.

Other marsupials include the tree-climbing koala and its cousin the lumbering burrow-digging wombat. There are many species of possum whose only relationship to the North American opossum is that they both spring from the first marsupials. They were called possums by the early explorers because they somewhat resemble their very distant cousin.

The possums range in size from the large odd-looking cuscus with its round head and tiny, close set ears to pigmy possums no bigger than small mice. There are also several species of marsupial gliders, resembling flying squirrels in their means of locomotion, and a marsupial anteater, the numbat.

The numbat, which lives in western Australia, is clothed in a beautiful reddish brown striped coat. Its special food is termites, and it has a long, slender snout and an even longer tongue with which it licks them up. It makes its home in the tree trunks, which the termites have conveniently hollowed out, and feeds on its hosts.

As for the kangaroos, theirs is a very large family. There are five species of large kangaroo—the eastern gray or forester, the western gray, the red, the antelope, and the stocky wallaroo.

Wallabies are simply smaller kangaroos, and there are many species of them—river wallabies, red-necked wallabies, pretty-faced wallabies, and rock wallabies, to name a few.

Still smaller kangaroos are known as pademelons. Among the pademelons are found the Tasmanian pademelon, the red-necked pademelon, and the red-legged pademelon.

One of the smaller members of the kangaroo family is the quokka, which isn't much larger than a cat. Then there is the hare wallaby, which gets its name because it darts out like a hare when flushed from hiding. The rat kangaroo looks like an oversized rat except that it has long feet and hops instead of runs.

Perhaps strangest of all the members of the kangaroo family is the stocky tree kangaroo, which hops on the ground but clambers up into the trees when it senses danger. Because it has to move along tree branches, it is the only kangaroo that has developed walking as well as hopping movements.

Though they may not look anything alike, all these marsupials are related. Scientists believe their common ancestor is the familiar North American opossum. As far as the records show, the opossum ancestors first appeared in North America in the Cretaceous period, around 100 million years ago. At that time, so scientists believe, Europe, Asia, and North America formed one massive landmass in the north while Australia, Antarctica, South America, and Africa formed another in the south.

As the Cretaceous period advanced, the landmasses began breaking up and drifting apart to form our modern continents. South and North America drew closer together, and the opossum was able to make its way southward, where it developed into many different species. Most of these South American species have since died out, while just one, the opossum, remains in North America.

Scientists speculate that some 60 million years ago the marsupial possums entered Antarctica and from there made their way to Australia. In those days the climate of Antarctica was warm enough to support fauna and flora. The landmasses had not become widely separated, so that a string of stepping-stone islands still connected the two continents.

However marsupial life began in Australia, fossils show that by the Oligocene-Eocene epoch, some 35 to 40 million years ago, it had become highly diversified. Strangely, no fossil remains of mammals dating back to this period have been found, though mammals had been developing in South America at the probable time of the marsupial migrations.

The first Australian marsupials were possums like those in

North and South America. At that time the Australian center was warm and humid, and rain forests flourished. One species of possum took advantage of the situation to come down out of the trees and develop into a large, hippopotamus-type marsupial. Another became a giant-sized koala, while a third emerged as a huge kangaroolike animal with a thick skull, short muzzle, and heavy teeth. A great carnivorous possum, the marsupial lion, haunted the caves of southern and eastern Australia. At the same time smaller marsupials such as the rat kangaroo were evolving.

After the last Ice Age, the climate started changing in Australia. Forests gave way to open savannahs. Much of the oversized fauna began to disappear. And the large kangaroos started developing.

Gradually, as Australia moved toward more modern times, the savannahs became increasingly dry, until the heart of the continent turned into desert land surrounded by semiarid steppes. The coasts remained damp and green, and there were subtropical conditions in the north. To adapt to their different situations, the kangaroo family began to diversify. By the time the western world learned of it late in the eighteenth century, it had evolved into a rich array of unique species.

When Captain James Cook brought his ship *Endeavor* into Botany Bay in 1770, he asked the aborigines who met him what the strange animal was called and was told, "Kangaroo." Many common names of other Australian animals were also given to them by the aborigines who lived here first. Scientists later classified the kangaroo family under the Latin name *Macropodidae,* which means "big foot."

Cook's exploration of the Australian continent was followed by an influx of new settlers who carved out farms and sheep and cattle ranches from the wilderness. Most of them looked on the kangaroos only as pests because they hopped in their fields or grazed in their paddocks. They began hunting them with dogs and guns and poisoning their water holes.

When bounties were placed on the heads of the kangaroos, professional hunters cruised the brush at night in their big trucks. Immobilizing the kangaroo with blinding spotlights, they would shoot them down at the rate of a thousand a week, moving on to the next locality when one was depleted. After the bounties were withdrawn, pet food processing plants and a growing worldwide demand for kangaroo pelts provided ready markets.

But here and there throughout Australia, people began waking up to the fact that they were letting a priceless heritage slip through their hands. The Kangaroo Protection Society was formed and began to campaign actively to save the kangaroo. Laws were passed in the various states to control the slaughter.

Though kangaroos are still being killed for pet food, restrictions now limit the number shot in each area. The harvesting of kangaroo pelts has become less lucrative since the United States government, under pressure by conservationists, banned their import.

However, despite laws and restrictions the slaughter still goes on, mainly by amateur hunters. Some are weekend poachers out for the sport of it. The kangaroo makes an easy mark because it will often just bound a few yards away and then turn to stare back curiously. It is this moment of hesitation that often costs it its life. Many of the poachers are not apprehended because areas in Australia are too extensive to be adequately patrolled.

In addition numerous kangaroos are killed accidentally when crossing roads at night. Transfixed by the oncoming glare of headlights, they are struck down by fast-moving traffic. Others succumb to poisoned bait or water holes, or break their necks or backs dashing against high wire fences.

So far the largest kangaroo species have managed to survive nationwide, though in some areas their numbers have been greatly diminished. In other areas they have increased, due in part to the water holes cattlemen have installed for their domestic stock.

But many of the wallabies, along with the other smaller marsupials, have not been so lucky. They suffer not only from the threat of predators but even more from the destruction of their habitats. As forests are cut down and underbrush cleared for farming, they lose both their shelter and their means of livelihood.

Even the change in Australian soil has pushed many of the small burrowers into extinction. When the first colonists came to the continent, the earth was so soft that their horses sank into it, sometimes up to their fetlocks. Australia at that time knew no hard-hoofed animals; today the imported cattle, sheep, and horses have pounded down the earth until it has become too hard for many of the smaller burrowers to penetrate. Deprived of shelter, they die.

What is the solution? As elsewhere in the world, the problem is the same—human beings as well as animals must live. And in areas where an overpopulation of kangaroos exists, some culling is necessary if the kangaroos themselves are not to suffer from starvation. Such supervised culling is presently being done by professional hunters whose accurate shots kill, not maim, and who make a careful selection, avoiding does with joeys.

A partial answer to the problem is to provide more spacious sanctuaries where wildlife can live unmolested. At present only two-and-a-half percent has been allocated for national parks in Australia, half the amount recommended by the United Nations as a minimum for conservation purposes. In the years ahead the government hopes to increase this amount to five percent. But many feel this is still not adequate.

Meanwhile, more and more private individuals have been stirred to pity at the sight of pouch young orphaned by fences, car wheels, and crackling guns. Dubbed "Marsupial Mothers" by some, they engage in the sometimes heartbreaking task of trying to raise the tiny victims that range from pouch joeys to the minute young of pigmy gliders. Too often, despite all their care, their attempts meet with failure.

In their quest for a greater tolerance toward the country's wildlife, Marsupial Mothers are appealing to the nation's youth. They visit classrooms accompanied by their small animal charges to tell their stories and introduce children and animals to one another.

An increasing number of educators, especially in the National Parks and Wild Life Service, are also attempting to give school children a deeper understanding and appreciation of the delicate ecological balance that sustains the unique fauna and flora of their areas. Competent instructors are conducting educational trips into the various reserves, including the Warrumbungles. Thousands of Australian school children take part yearly in such back-to-nature programs.

Children have always been the hope of conservationists around the world, for they have discovered there is no limit to what a child can accomplish once aroused. In America a fervent campaign conducted by children saved the wild mustang from extermination when everything else had failed. It may well be that the informed children of Australia will spearhead the growing conservationist movement in their country.

# Bibliography

*Books*

BARRETT, CHARLES. *Wild Life in Australia*. Australia.

BERGOMINI, DAVID. *The Land and Wildlife of Australia*. New York, 1964.

BREEDEN, STANLEY and KAY. *The Life of the Kangaroo*. New York, 1966.

————. *Wildlife of Eastern Australia*. New York, 1973.

BUSTARD, ROBERT. *Australian Lizards*. Sydney, 1970.

CARTER, JEFF. *Wild Animal Farm*. Australia, 1972.

DAVEY, IRENE. *Australia, Colourful Continent*. Australia, 1972.

DAVEY, KEITH. *Australian Marsupials*. Melbourne, 1970.

FRITH, H. J. and CALABY, J. H. *Kangaroos*. Melbourne, 1969.

GOULD, JOHN. *Kangaroos*. With modern commentaries by Joan M. Dixon. 1973.

GUILFOYLE, W. R. *Australian Plants*. Australia.

HALL, JOHNSON and CHIPPENDALE. *Forest Trees of Australia*. Canberra, 1970.

HARRIS, THISTLE Y. *Alpine Plants of Australia*. Sydney, 1970.

HILL, ROBIN. *The Corner*. Lansdowne, 1970.

KELLY, STAN. *Eucalypts*. Text by G. M. Chippendale and R. D. Johnston. Australia, 1969.

KNOWLES, PETER. *Australia's Wild Heart*. Australia, 1972.

MARLOW, B. J. *Marsupials of Australia*. Australia, 1962.

MILLETT, MERVYN. *Native Trees of Australia*. Melbourne, 1972.

MULLINS, BARBARA and MARTIN, MARGARET. *Warrumbungle National Park.* Australia.

NATIONAL PARKS AND WILDLIFE SERVICE. *Warrumbungles National Park.* New South Wales, 1972.

RIDE, W. D. L. *A Guide to the Native Mammals of Australia.* London, 1970.

SERVENTY, CAROL and HARRIS, ALWEN. *Rolf's Walkabout.* Australia, 1976.

SERVENTY, VINCENT. *Wild Life of Australia.* New York, 1972.

SLATER, PETER. *A Field Guide to Australian Birds.* Vols. 1 and 2. Australia, 1970.

TROUGHTON, ELLIS. *Furred Animals of Australia.* Revised edition. Sydney, 1973.

TYNDALE-BISCOE, C. H. *Life of Marsupials.* Australia, 1962.

WORRELL, ERIC. *Reptiles of Australia.* Sydney, 1963.

## Scientific Papers and Periodicals

ANONYMOUS. "Of Time and Deeds and Empty Cities." *Australian Natural History,* Special Supplement, September, 1976.

———. "Kangaroos and Men." *The Australian Zoologist,* 1971.

BROWN, G. D. and MAIN, A. R. "Studies in Marsupial Nutrition." *Australian Journal of Zoology,* 1967.

CAUGHLEY, GRAEME. *A Census of Kangaroos.* University of Sydney, 1976.

———. "Social Organization and Daily Activity of the Red Kangaroo and the Grey Kangaroo." *Journal of Mammalogy,* 1964.

CHILDREN'S MEDICAL RESEARCH FOUNDATION. *Director's Report, 1973–74, Regarding Cataracts in Kangaroos.* Sydney.

CLARK, M. J. and POOLE, W. E. "Reproductive System and Embryonic Diapause in the Female Grey Kangaroo." *Australian Journal of Zoology,* 1967.

DIVISION OF WILDLIFE RESEARCH, LYNEHAM, A. C. T. *Kangaroos.* Includes the following articles:

"Sexual Maturity and Oestros," by W. E. Poole and P. C. Catling, 1974.

"Gestation, Parturition, and Pouch Life," 1975.

EALEY, E. H. M. *Ecology of the Euro in North-western Australia.* Series No. IV. *Age and Growth.* CSIRO Wildlife Research, Australia, 1967.

FOX, ALLAN M. "Warrumbungle Mountains." *Australian Natural History,* Vol. 18, No. 88.

GOLDSTEIN, WENDY. *Koala.* National Parks and Wildlife Service, Sydney.

GRANT, T. R. "Dominance and Association among Members of a Captive and a Free Ranging Group of Grey Kangaroos." *Animal Behavior,* 1973.

KIRKPATRICK, T. H. "The Grey Kangaroo in Queensland." *Queensland Agricultural Journal,* 1970.

————. *Studies of Macropodidae in Queensland.* Series of articles from *Queensland Journal of Agricultural and Animal Sciences,* including:
"Food Preferences of the Grey Kangaroo," 1965.
"Age Estimation in the Grey Kangaroo, the Red Kangaroo, the Eastern Wallaroo, and the Red-necked Wallaby," 1965
"Reproduction in the Grey Kangaroo in South Queensland," 1965
"Social Organization of the Grey Kangaroo," 1966
"Effects of Drought on Reproduction in the Grey Kangaroo," 1966.

KIRKPATRICK, T. H. and MCDOUGALL, W. A. "The Grey and Red Kangaroo in Queensland." *Australian Zoologist,* 1971.

MORRIS, ALAN. "Wildlife Column." *Coonabarabran Times,* 1976. Includes the following articles:
"The Birds of the Warrumbungle National Park"
"The Cattle Egret"
"Return of the Migrants"
"Wattles of the Castlereagh Region"
"Where Do Kingfishers Go in Winter"
"White-eared Honeyeaters"
"Wildlife Notes" (Survey of birds in Warrumbungles area).

MORRIS, A. K. and FOX, A. M. *The Wedge-tailed Eagle.* National Parks and Wildlife Service.

NATIONAL PARKS AND WILDLIFE SERVICE, Sydney, N. S. W. *Notes on Australia's Unique Mammals.*

————. *The Conservationist.* 1976.

————. *Parks and Wildlife.* December, 1974.

————. *Parks and Wildlife*. Vol. 1, No. 5.

————. *The Third Alternative*.

————. *Wildlife Refugees*. June, 1976.

POOLE, W. E. "Breeding, Biology and Current Status of the Grey Kangaroo of Kangaroo Island." *Australian Journal of Zoology*, 1976.

————. "The Eastern Grey Kangaroo, Micropus giganteus in Southeast South Australia." Division of Wildlife Research, 1977.

————. "A Study of Breeding in Grey Kangaroos in Central New South Wales." *Australian Journal of Zoology*, 1973.

POOLE, W. E. and CATLING, P. C. "Kangaroos" I. Sexual Maturity and Oestrus, 1974. II. Gestation, Parturition and Pouch Life, 1975. Division of Wildlife Research, Lyneham.

POOLE, W. E. and PILTON, P. E. "Reproduction in the Grey Kangaroo in Captivity." CSIRO Wildlife Research, 1964.

RUSSELL, ELEANOR M. *The Biology of Kangaroos*. University of New South Wales, Kensington, Australia. Reprinted from *Mammal Review*, 1974.

SHARMAN, C. B. and CALABY, J. H. "Reproductive Behavior in the Red Kangaroo in Captivity." CSIRO Wildlife Research, 1964.

STEPHENS, TANYA. "Nutrition of Orphan Marsupials." *Australian Veterinary Journal*, 1975.

STEPHENS, TANYA; CROLLINI, C.; MUTTON, PATRICIA; GUPTA, J. D.; and HARLEY, J. D. "Galactose Metabolism in Relation to Cataract Formation in Marsupials." *Australian Journal of Experimental Biology and Medical Science*, 1975.

STEPHENS, TANYA; IRVINE, SUSAN; MUTTON, PATRICIA; GUPTA, J. D.; and HARLEY, J. D. "The Case of the Cataractous Kangaroo." *Medical Journal of Australia*, 1974.

————. "Deficiency of Two Enzymes of Galactose Metabolism in Kangaroos." *Nature*, 1974.

WARRUMBUNGLE NATIONAL PARK. *Checklist of Birds*.

————. *Standard Mammal Checklist*.

*Interviews*

Lynn Alva, assistant to Eric Worrell, Australian Reptile Park, Gosford, Australia.

Dick Duggan, chief ranger, Warrumbungles National Park.

Mr. and Mrs. Keith Fitzgerald, marsupial mothers.

Allan Fox, Environmental Education and Wildlife Extension Service, New South Wales National Parks and Wildlife Service.

Dr. H. J. Frith, Chief of Division of Wildlife Research, CSIRO, Australia.

Peter Krauss, wild animal collector for zoos, Queensland, Australia.

Basil Marlow, Curator of Mammals, the Australian Museum, Sydney, Australia.

Chris Martin, assistant ranger, Warrumbungles National Park.

Alan K. Morris, senior ranger, Coonabarabran District.

William E. Poole, Division of Wildlife Research, CSIRO, Australia.

Mr. and Mrs. Raymond Raisbeck, curators, Blackbutt Reserve, Kotara, New South Wales; marsupial mothers.

Don Scots, assistant to Graeme Caughley, School of Biological Sciences, University of Sydney, Australia.

Dr. Tanya Stephens, B. V. Sc., marsupial mother.

# The Snow Monkey at Home

For Colleen

## ACKNOWLEDGMENTS

I would like to express my gratitude to my many new friends in Japan who went out of their way to help me in this project. Special mention should be made of Dr. Kinji Imanishi, Dr. Naoki Koyama, Dr. Noboru Nakamura, Dr. Akio Mori, Prof. Shigeru Azuma, Shigetaka Kotera, Eishi Tokita, Chikao Nakata, Satsue Mito, Kazuhiko Ishikawa, Tokui Aiba, and Edward Hains.

# Foreword

Throughout Japan scientists have opened many primate research centers where they provide food for the macaque troops so that they can study their ways. One of these centers is the Jigokudani center situated near a spa in the Japanese alps. The snow monkeys who are drawn to the center with the promise of food are designated by the researchers as Troop A.

This book is not about this particular troop. It does, however, draw heavily on the scientists' research conducted here and at other monkey centers to trace the activities of another more isolated troop that lives in the general vicinity.

It should be understood that in describing the social organization of macaque monkeys, we are using the term "center" and "periphery" in the sense they are used by the scientists. According to Dr. Akio Mori of the Kyoto University Primate Research Center at Koshima, these terms do not constitute a strict geometrical pattern in which the monkeys arrange themselves. Rather, "center" and "periphery" should be understood in the same sense as "inner circle" and "outer circle" would be applied to human relationships. They are simply convenient labels to distinguish between the various rankings and divisions of power in a monkey troop.

Scientists have ascertained such rankings by throwing food between two monkeys. Only one, the dominant monkey, will reach for it in the presence of the other. Other dominance behaviors are performed by mounting—the dominant monkey mounts the subordinate one—by one-sided attacks—the attacked monkey does not retaliate—and by similar behavioral patterns. By noting these behaviors the scientists have been able to lay out a rather orderly pattern of the social hierarchy among Japanese macaques.

# Contents

# The Snow Monkey at Home

# Prologue

The Japanese islands, among the world's newest land masses, are part of a long ridge of mountains that rise from the ocean floor, stretching from Southeast Asia to the Aleutian Islands. Four of these islands—Hokkaido, Honshu, Shikoku and Kyushu—together with many smaller ones make up Japan, a country off the east coast of Asia.

Almost three quarters of Japan's land mass consists of steep hills and perpendicular mountains whose slopes are covered with forests into which farmers and lumberers have made deep inroads. In the remnants of these forests dwells a unique member of the monkey family, the Japanese macaque, or *Macaca fuscata*. Living at a latitude of 35 to 45 degrees above the equator, the *Macaca fuscata* is the most northern of the monkey species. It is indigenous only to Japan and is scattered through all the islands except Hokkaido.

Because the Japanese islands are situated longitudinally, they have extreme variations of climate. Kyushu and Shikoku, in the south, seldom see snow, and most of their trees are evergreens and leafy the year round. Hokkaido, at the northern end, has long, bitter winters. The largest island is Honshu, which the Japanese refer to as their mainland. Honshu is shaped like a bent bow; its northern tip shares Hokkaido's climate, and its southern tip is as temperate as Kyushu and Shikoku. The island's backbone is made up of steep chains of mountains that culminate at the center in the Japanese Alps, where the highest peaks reach close to 10,000 feet above sea level.

The Japanese monkey is found from northern Honshu to southern Kyushu—even in the wintry solitudes of the Japanese Alps. Here the temperature in winter often falls below freezing, the days are short, and the trees are deciduous, losing their leaves in the fall and remaining barren for five months of the year.

The monkeys that live in the northern province and in the Alps are known as snow monkeys. They wear woolly winter coats, pale gray in color. Snow monkeys are larger than the general run of Japanese macaques, but the size of their troops is much smaller. The monkeys that live in the more temperate zones have brownish coats. Their troops may number more than two hundred individuals in some instances. There is also a subspecies, the *Macaca fuscata yaqui,* which consists of a single troop that lives on the small offshore island of Yakushima. This monkey has greenish fur and a flatter face than the other macaques.

Though all Japanese monkey troops have a similar culture, there are many minor differences among them. The vocalizations by which monkeys communicate within their troop vary slightly one from another. Each troop has also developed its own food habits. What is staple macaque food in one locality may be shunned in another as if it were poison.

There are other cultural differences. For example, the monkeys of

Takasakiyama, in northern Kyushu, are the biggest scrappers in the islands. During the mating season the males conduct their courtship by attacking the females so brutally that they are covered with blood. Nevertheless, a tender side to the Takasakiyama males appears during the birthing season. At this time, high-ranking males care temporarily for the yearlings of females that are bearing another child. This care is seldom seen in other troops and is quite different from the occasional adoption of an infant by a male, which is found in many troops.

Sometimes new habits are formed in different troops, introduced by curious youngsters. At Koshima, in southern Kyushu, a young female started washing sweet potatoes before eating them. She also learned how to separate wheat grains from sand by tossing a handful of sand and grain into a tidal pool. The sand sank and the clever monkey scooped up the grains, which floated to the surface. In the Japanese Alps, snow monkeys have learned to wash apples that have been sprayed with insecticide before eating them. And the monkeys of Arashiyama, near Kyoto in Honshu, have come to savor eggs so much that they have become inveterate nest robbers. No bird builds a nest in the vicinity of these monkeys' trails.

Among all the troops, the snow monkeys of Shiga Heights, in the Japanese Alps, stand out because they have been able to adapt to a climate so bitter that few other macaques could survive in it. Five troops live in this area, which is located high in the mountains of Nagano Prefecture. Each troop has its own territory in a region that varies from 2600 to 6300 feet above sea level. The landscape is one of heavily forested mountain slopes, deep ravines, and rushing mountain torrents. Hot sulphur springs scattered throughout reveal the mountains' volcanic origins.

Spas have been developed around some of the larger springs. And visitors come here for stimulating medicinal baths. In the winter when the mountains are clothed with snow, other visitors arrive from all over Japan and from foreign countries to ski.

But the visitors interfere little with the life of the wilderness creatures in these alpine forests—with the snow monkeys, the serow (the Japanese antelope-goat), the Asiatic bear, the raccoon dog, the marten, the hare and the fox. They live out their lives in the fastnesses apart from most human beings.

# The Birthing

In the predawn gloom the forests that clothe the steep slopes of the Japanese Alps lie hushed, waiting for the moment when daybreak will call out responses of birdsong and the rustlings and chatter of animals. It is late May—the quickening time of the year. Spring has come in earnest to these slopes more than five thousand feet above sea level. Only on the high, barren peaks do remnants of winter's snow cling stubbornly.

At their night campgrounds in the branches of the cryptomerias most of the troop of snow monkeys are still sleeping. Their pink faces and buttocks, framed in long gray hair, glimmer faintly under the lightening sky. Only two are awake and stirring—a young female monkey and her mother. They have spent the night huddled together on a high branch.

The young female is expecting her firstborn, conceived during the mating season some six months before, when she became sexually mature.

Among the Japanese monkeys the mother-daughter relationship is very close throughout life. The bond seldom loosens. Although this mother can do nothing to help with the birthing, she will stay close by, her presence perhaps a comfort to her daughter.

The young female is only five years old and will not reach her full growth for another two or three years. Somewhat smaller than her mother, she is less than two feet long from the top of her head to her buttocks. In the last few weeks she has lost weight because her appetite has lessened, but she still has a sleek, well-fed look. Her swollen abdomen makes her seem larger than she actually is. Also, she will not shed her thick coat of gray winter hair until after her baby's birth.

Now that the young female has felt the first faint birth pangs, she knows she must be on her way. The two agile monkeys slip along the branch of the cryptomeria to its trunk. They do not have the help of the long tails that distinguish arboreal monkeys. Snow monkeys discarded their long tails ages ago when they took to ground living. Now their short, stubby ones, barely six inches in length and tufted at the end, are of no use whatever in climbing.

But the monkeys still spend a great deal of time in trees and have no problem getting around in them, for their strong hind feet with their widely spread big toes enable them to cling to branches, upside down if need be. And their front limbs, with hands that possess four long fingers and an opposing thumb, are also strong. Their deft fingers can grasp the smallest branch or pick up the tiniest objects with ease.

The two monkeys slide down the trunk buttocks first, and quickly disappear into the undergrowth below. Usually, birthings among Japanese monkeys take place in the dark hours. When the time is upon them, the females search out a secluded spot in which to bear their offspring, away from all observers except, perhaps, a female relative—a mother, a sister, a daughter.

The pregnant monkey makes her way to a low outcropping of rock and sits with her back against a boulder, her mother squatting beside her. As the birth pangs quicken, mucus pours from the birth channel. The young female stumbles up and bends over the rock, letting it support her. Eyes closed, mouth agape in distress, the little female pants in quick gasps as she strains to give birth.

Her labor will be easier than that of human mothers, however, for it will be much shorter. This is true because the weight of the baby is borne by the backbone in creatures that walk on all fours, whereas in human beings the weight is carried in the pelvic region. Pelvic bones and muscles developed to hold the embryo in place make the birth channel narrower so that birthing often becomes a long and difficult process for two-legged beings.

Half an hour after the first birth pangs began, the tiny head of the infant monkey appears. Bent over, the new mother clasps the head in her hands and pulls. In less than a minute she has drawn the whole body out, followed by the afterbirth and umbilical cord to which the baby is attached.

The young female stares at the tiny blood-covered infant, which is sprawled out on the ground before her. Then she turns her attention to the afterbirth. Like most mammals, she begins eating it, pausing now and then to lick the blood from the baby's matted fur or to cleanse herself. As she eats the afterbirth she bites through the umbilical cord, leaving a length of some ten inches trailing from the baby's belly button.

With the afterbirth gone and the rock where the birthing took place licked clean of all blood, the snow monkey lifts up her tiny infant. The brightening sky shows a toothpick-thin creature, scarcely four inches long from head to buttocks. Its tiny face is flushed a bright red. The baby is fragile in its mother's arms, weighing less than a pound. A fine, soft mat of dark baby fur covers it.

Carefully the mother examines it—face, ears, spider hands and feet, minute buttocks, and genitals. It is a male, perfect in every detail. It lies in her arms still wet from its birth while she continues to lick it, pausing now and then to bend over and peer at it with large, round yellowish hazel eyes. Those eyes seem to be mirroring unanswerable questions: What is it? Where did it come from?

The baby's eyes, closed at first, presently open and return her gaze with an unfocused stare. One day, sight will be this monkey's most valuable sense organ. For the gift of acute vision, his species has sacrificed the sense of smell. Unlike the dog and other long-muzzled creatures, the monkey's smelling sense is not acute. In the flat face the nose is small with close-set down-turned nostrils. But the eyes, placed on either side of the face like human eyes, will one day have not only sharp vision but also will be capable of judging perspective and distinguishing colors.

Hearing, too, is good in the macaque, though the ears are small and stationary, unlike the long flexible ears of the hare, which can swivel in any direction. Instead, like all monkeys, the baby must learn to turn his whole head to locate and identify what he hears.

Now the dawn symphony of the wakening forest pours into the ears of the tiny baby—the stirring of the breeze, the querying *ku ku ku* of awakening monkeys, the murmuring of wood pigeons. But closer than wind or animal or bird sounds are the soft sighings and rustlings of his mother's breathing, in and out, in and out, and the feel of her warm breasts rising, falling. Those sounds, and the sense of warmth in the arms that enfold him, will make up the narrow horizons of the baby's early days. That and food—suddenly he is ravenously hungry.

*Eh eh eh,* he murmurs in a voice almost too weak to be heard. With his tiny mouth he gropes against his mother's soft white breast searching for one of her two nipples. Finally he locates the protuberant pink tube and fastens his lips on it. Silent now, he nurses greedily.

The tiny hands, surprisingly strong even at birth, clutch at the mother's thick fur. The strength in those small hands and arms will enable the baby to survive and reach maturity. For, unlike many animals, Japanese monkeys have no lair in which the young may be left while their parents scour the land for food. Macaques lead a nomadic life, and babies must travel with the troop.

Now the monkey voices are louder—infants squealing, adults calling to one another as they gather in the clearing beneath the cryptomeria trees. Presently they will be on their way in search of the day's food. The little female's mother has already disappeared. After this brief rest she, too, must be on her way.

One arm clasping her baby tightly against her breast, she sets off in a three-legged gait for the clearing. Disturbed by the motion, the baby loses the nipple and murmurs the tiny complaints, *eh eh eh,* against her breast. But the mother ignores it, and soon the greedy little mouth finds the nipple again.

A new life has been born, and a different way of living has opened for the young mother. She has passed from the threshold that separates the juvenile from the adult. This change has taken place in little less than an hour from the time she left her treetop perch.

# 2

# A Trek Through the Forest

$B$y the time the female monkey enters the clearing, the others have descended from their night perches. Not counting this season's babies, there are some thirty monkeys in the troop, seventeen of them females. As troops go it is a small one, especially compared with those in warmer parts of Japan, some of which number more than two hundred. But in these Alps, where snows are heavy and winters are lean and bitter, smaller troops have a better chance of surviving. So when the troop becomes too big it divides into two parts which go their separate ways.

A few juveniles are wrestling playfully together. Some adults are feeding in the trees, while others are scattered in small groups whose members are grooming one another. Ordinarily, Japanese monkeys cannot bear to be in close bodily contact. A mere touch on the skin by another monkey may be construed as a challenge to fight. Grooming is the only social activity that brings monkeys together. Even the old leader of the troop, who seldom permits anyone to approach within five feet of him, enjoys being groomed. The new mother sees a young female squatted over his prostrate body deftly parting his grizzled hair, turning it sideways layer by layer as she picks through it. Specks of dandruff, tiny twigs, lice, scabs—whatever she removes she places in her mouth and chews two or three times while continuing her grooming.

The sight of the old leader, eyes closed, a look of contentment on his blotched face, discolored by age, fills the young mother with a longing for that same pleasant sensation of close contact. When she sees her mother, her older sister, and her grandmother huddled together grooming one another, she moves toward them tentatively.

As she approaches, all three turn in unison to gaze at her coldly. A stare, like a touch, is an unfriendly gesture. Daunted, she hesitates, then slowly continues her approach. She stops short again when her older sister, still staring at her, lowers her head, wrinkles her forehead until the hair and ears lie straight back, and thrusts out her mouth, making soft but explosive pufflike sounds. The mobility of the monkey's lips, which are not anchored to its gums as in most other animals, allows it to express its feelings visually as well as orally. The young mother plainly understands that she is being rejected by her family. No matter how close the bonds of kinship may be among Japanese female macaques, this rejection of a newborn infant and its mother seems to be the customary reaction.

Now the young mother backs away from her hostile family. But her yearning to groom and be groomed grows even stronger. Not far off she sees another huddle of groomers, this one made up of two young mothers with week-old babies and two other mothers with year-old juveniles—a nursery group. She edges toward them.

The one-year-olds in the huddle spot the newborn baby and crane their necks curiously. Then they gallop toward the young mother for a closer investigation. They don't get far. One mother grabs her child by its leg, and the other holds hers by the tail. Both are hauled back to the group, screeching loudly.

*A Trek Through the Forest* : *181*

As the young mother continues in their direction the whole group withdraws silently but swiftly. She is alone again. Bewildered, she retreats to a corner of the clearing. As she goes, monkeys in whatever groups she nears melt away from her.

Males seem particularly fearful of the tiny infant. Even the old leader, who inspires such respect—almost fear—among the other monkeys, jumps to his feet and backs away when he sights the baby. And the female who has been grooming the leader runs off.

Finally the young mother reaches the secluded nook to which she has been withdrawing. She sits down and turns all her attention to the baby. She begins to reexamine him. Then, grabbing him by the tail, she turns him upside down and starts licking and cleaning his buttocks once more.

The baby loses the nipple and begins his peevish *eh eh eh,* feverishly rooting over her breast until she allows him to nurse again. While he nurses she grooms him for a few seconds, picking through the soft mat of baby fur.

Then, unable to leave him in peace, she grasps the hair above his forehead and bends his head back so that the nipple is again jerked from his mouth. Ignoring his protests, she peers intently into the strange tiny face of this newcomer who, for some unknown reason, has turned her into an outcast. Finally she releases the baby and begins instead to groom herself, scratching at her thigh and occasionally pausing to stare straight ahead, rocking back and forth slightly and crooning *ku ku ku* to herself.

The leader is now stalking about the clearing. As he does so, the other monkeys fall away from him. He has undisputed authority over this troop, even though he is now more than twenty-five years old and blind in one eye. Other males are stronger and larger than he. But the leader has the confidence of the troop, especially of the females, who play a decisive role in the selection of leaders.

For many years, this leader has fulfilled his duties well. He patrols resting and sleeping places and stops quarrels, often just by ambling over to the brawlers with a threatening grimace. He also guards females and their young during the troop's nomadic wanderings. Though his aggressiveness has been fading with age, he has two subleaders on whom he can rely to help him with these chores.

Now he will select the trail for the day's foraging. There are a number of these trails, worn by years of usage. They meander over the troop's territory, which has a diameter of some three miles. The monkeys seldom pass beyond the limits of their territory with its sharp mountain ridges, steep slopes and narrow ravines through which torrents pour. There is some open meadowland, but most of the territory is wooded because monkeys are forest dwellers.

The monkeys do not cover the whole area at one time. Each season has its particular locality where food supplies are richest. Trails chosen by the leader pass through these localities as the month dictates.

Some of the monkeys' treks are short and swift. Then they usually take to the trees, swinging easily from limb to interlacing limb. But this is to be a long, leisurely ramble on ground level with frequent rest periods and foraging along the way.

When the leader selects the trail he calls out, *Kwaa! Kwaa!*

The chief female, a fierce old grandmother, answers, *Hui! Hui!*

The cry is taken up and swells from all directions as monkey after monkey echoes it.

*Hui! Hui! Hui! Hui!* The troop is on its way.

The younger, more adventuresome males soon outstrip the leader. They go at their own pace, scattering now and then to explore the surrounding area, chasing one another, pausing to play and then galloping on ahead. But if they go so quickly that they cannot hear the *hui hui hui* behind them, they stop and turn back until they have come again within sight of the leader.

Accompanied by the subleaders—one in front, the other behind—the leader moves forward with stately steps. He is surrounded by the females and their young. Some are carrying small babies clinging to their breasts. Older youngsters ride astride their mothers' backs or trudge at their sides.

The new mother, still clasping her baby with one arm, stumbles along on her three-legged gait. Her infant's hands and feet cling tightly to her fur as he suckles. His mother does not yet dare trust those weak limbs to hold his full weight.

The rest of the juveniles and young male adults bring up the rear. They are as adventuresome as the front-runners. Some of them, knowing the goal of the trek, break off from the troop to take a shortcut, swinging through the trees to arrive at the first stopping place ahead of everyone else. Others straggle into the wilderness on curiosity-satisfying missions of their own. It is up to the subleaders to see that all arrive safely at the troop's destination.

It is a time for kicking up one's heels. All about, the kaleidoscope of spring unfolds. New green leaves sparkle in the sunlight. Catkins sway on willows and oaks and poplars. Tiny clusters of greenish flowers dance in the maple trees.

Food is everywhere. The red knobs of ripening wild strawberries wink from the secret shadows of their leaves. Vines of the wild yam tangle in damp hollows, where the monkeys greedily eat their leaves and buds, even though they have not learned to dig in the ground for the edible tubers. Knotty wisteria, rich now with white and lavender tassels, and wild grape vines twined round the trunks of tall trees, also provide tasty young leaves.

In open glades new grass and clover, studded with white blooms,

grow rich and green. Yellow and pink violets and feathery pink dianthus nod in the shadowy places. Marshy lands gleam with golden marigold, while overhead the white flowers of wild magnolia fill the air with fragrance.

Above the flowers, huge black butterflies soar on lazy wings, defying the monkeys' efforts to grasp them. Gold-tailed bumblebees buzz officiously through sun-patterned tree corridors. Dragonflies flit over pools where the monkeys, pausing to drink, lift the water up with cupped hands or bend over to lap at it.

Now and then there is the flash of a blue flycatcher on the wing. Gaily colored woodpeckers perched on dead branches or hollow tree trunks converse with one another, their *rat-a-tat-tats* drummed out by hard beaks. A black crow, chased by a swarm of irate swallows whose nests it had been marauding, flaps off among the trees, cawing dismally. And over all fly the great tawny kites with wingspans of three feet, crying *pee-oui, pee-oui* as they scan the ground for carrion. Live squirrels, hares, and other small forest dwellers must run from the kites, though the monkeys have little to fear from them.

The *hui hui* of the monkeys punctuates the other forest sounds. It is their means of keeping in touch with one another, for they do not maintain a tight formation. The first scout to reach the stopping place calls out a triumphant *Hogaa! Hogaa!* This quickens the steps of the others.

Once the monkeys arrive at their destination they spread out in all directions to feed. Occasionally, *Hogaa! Hogaa!* will sound again. A monkey has wandered so far away it can no longer hear the calls of its companions and is sending out an urgent query. Moving this way and that, it keeps up the cry until it hears an answer and can reorient itself.

Stopping often throughout the day, the monkeys complete their leisurely trek by late afternoon, when they reach the night's campsite. At

*Forest companions: red fox and racoon dog*

every stop the new mother has been trying desperately to join a group of groomers, but each time she has been firmly, even angrily, rebuffed. Now, as the sun draws near the western horizon, she finds herself still alone, shunned even by her family.

Sounds of contentment are all around her—the soft *ku ku ku* of monkeys grooming or asking to be groomed. In sudden frustration she seizes her suckling baby by the shoulders and shakes him. He screeches his protest, a thin, wavering cry like that of a night bird. She clasps him to her, and patting him soothingly on the back, lets him suckle.

# ③
# The First Month

By the end of the second day, the baby's beet-red face has paled to an amber color. The umbilical cord, which still hangs from its belly like a ribbon, has begun to shrivel. Soon it will drop off. The huge eyes of the baby stare wisely from a tiny, gnomelike face.

The young mother, who has never stopped begging, is at last admitted grudgingly into the huddle of her family. Once this happens, she and her baby are accepted everywhere by the females. The males, however, will continue to give her baby a wide berth until he is at least a month old. They still act almost as though they are afraid of the tiny creature.

As the days go by the baby becomes an irresistible attraction for the females, especially those who have borne no young this year. In their eagerness to get close to babies, childless females try to insert themselves into nursery huddles. Giving long, low *kus* and moving their lips and scraping their teeth together, they back timidly into the magic circle, fearful of being driven away. Even when they are admitted, it is difficult for them to touch a baby because the mothers will not let anyone, even their own relatives, handle their young.

Grooming is the only way to disarm the mother. But grooming, which requires such close physical contact, is not entered into lightly by the monkeys. It must be preceded by a kind of ritual.

The one wishing to groom says softly, *ku ku ku ku,* while looking inquiringly into the face of another monkey. If the other replies with a smacking of the lips and then lies down, the offer is accepted. But it is understood that the one being groomed will then return the favor.

If a monkey wishes to request a grooming instead of giving one, it simply makes eye contact with the one it has chosen and then stretches out in front of the prospective groomer on belly or side, or sits with head bowed or body bent backward. All are grooming postures. The monkey wishing to be groomed may also smack its lips. It is then up to the other to decide whether it wishes to groom or not.

More groomings take place between female adults, or between females and males, than between males alone. But most groomings occur with mothers and their offspring, where affection takes the place of ritual and few signals are required. Sometimes three members of a family form a huddle of mutual groomers.

It is not surprising that the young mother accepts without question her grandmother's offer to groom her. Stretched out on her side, her baby clasped to her breast, she gives herself up to the delicious sensation. But, soon after she closes her eyes, she feels a sharp jerk on her baby's leg and sits up abruptly. Her grandmother has hold of the leg and is pulling the baby out of his mother's arms in an attempt to kidnap him. Before the older monkey can make her getaway, the baby's mother grabs a tiny arm. Now a tug-of-war begins, with the baby stretched between the two monkeys like a limp rag, unable to utter a cry.

If the grandmother is successful, she will never let the mother come near enough to take the baby back again. And though the grandmother will cradle him in her arms, croon over him, and lavish love on him, the baby will die in a week or two because she will have no more milk to feed him.

At last the young mother gives a final desperate jerk and pulls her baby free. Hugging him tightly, she makes threatening noises at her grandmother, her lips thrust out in a rounded O. The grandmother creeps away, her head bowed.

Presently another female comes to offer her grooming services, and her offer is accepted. She squats behind the mother, assiduously raking through her fur until her eyes close. Then quickly she reaches around to touch the baby's head. He squirms and mutters *kyot*, and the mother's eyes fly open. The groomer jerks back her hand and begins to groom again. In a few minutes she will try once more.

The young mother's new responsibilities have been a drain on her physically. Nursing the baby, as well as her inability to forage properly, have caused her to lose weight steadily. But things become a little easier in the next few days as she can trust the baby to hang on to her belly without the need of her supporting arm. This enables her to walk on four legs again.

She also is able to put the baby down for short periods while she forages, but she is careful to choose some out-of-the-way nook in which to lay him, safe from the trampling feet of the other monkeys. And she never goes far from him. While he sleeps she gathers food, storing it in her cheek pouches until they are distended like small balloons. She will take time to eat after returning to her baby. Her ears are always alert for any sound from him.

Whenever he wakes and finds her gone he utters a new cry, a mother-calling cry used by all infant Japanese monkeys: *hui hui hui*. It is a soft call that can scarcely be heard a few feet away. But the mother always responds to it and hurries to the baby.

She picks him up and cuddles him against her breast, where he gropes for a nipple and begins to nurse, while she eats the food stored in her cheeks. She carefully grooms out any particles that may fall on the baby's fur, the grooming itself a caress. Finally full, he lies there murmuring, *kuor kuor ku*.

He is always contented in his mother's arms. To him the whole world is made up of her presence, her soft fur, her solicitous pink face with the large round eyes looking down at him. In her arms he experiences safety and contentment. Even when he is not nursing he likes to keep the nipple in his mouth.

He cannot bear being separated from her. When he finds himself slipping from his mother's body or when his mother moves suddenly, frightening him, he lets out a furious *kiiii* or *kyot*, twisting his head violently from side to side. He has a different cry when he cannot find the nipple, chattering *k.k.k.k*. And when he murmurs in his sleep it sounds like *ke-e-at*.

He does not spend as much time in babyhood as do human children. His growth is rapid. On the second day of his life his first milk tooth erupts. During the following days others pop out one by one so that by the time he is three weeks old he will have all his front baby teeth.

Within a week he is crawling around his mother. By the end of the week he begins scratching his thighs when he wakes from a nap. It is the first sign of the grooming instinct. Another wakening instinct is climbing. Almost from birth he has been trying to pull himself up on anything his hands touch. Laboriously he hauls himself over his mother's thighs and up her back. One day he crawls to the trunk of the bush under which he and his mother are resting.

Up, up, up he begins to climb. When he reaches the first branch he swings out on it and hangs there like a rubber toy. He is not yet strong enough to haul himself up on it, but he is afraid to let go and drop down.

Locking his elbows convulsively he shrills, *kyot kyot kyot*. His mother lifts up her arms and brings him down, patting him on the back to comfort him. Shortly afterwards he clambers up again, and again, hanging stranded from the branch, cries out for help.

This time, instead of taking him in her arms, his mother stretches up to put her face on the same level as his. Looking straight at him, she moves her lips in silent encouragement, in and out, in and out, in a kind of rhythm. When he stops crying, she begins grooming him on his back and legs. Finally he realizes there is nothing to fear, then she lifts him down again and lets him suckle.

Day after day he practices his climbing, until his arms are strong enough to pull himself up to the branch from which he is hanging. So far he has not tried to descend by himself, but one day he starts down the trunk of the bush upside down. Partway he turns himself around, pivoting carefully and slowly, his tiny clawed hands digging into the bark, and descends triumphantly rump first.

As the baby becomes larger he is threatened with another kind of kidnapping. Juvenile female monkeys who are nearing the threshold of adulthood like to play house with baby monkeys. One day, as the baby is crawling about his dozing mother, his two-and-a-half-year-old cousin grabs him and carries him off to a nearby sapling. She raises him to a low branch and makes him hang on it, higher than he has ever been before. Then, standing behind him on her hind legs, she begins performing rhythmic lip movements at him while he clings meekly there.

Finally his mother comes to his rescue. She threatens the young female with a fierce grimace and yaps at her loudly. It is such a severe scolding that the guilty juvenile scuttles off, but she does not go far. A short distance away, she stops and waits for another chance.

The baby's eyesight has been developing rapidly. Now he can focus, which enables him to take a keen interest in objects around him. On his thirteenth day he begins picking up twigs and playing with them. Hard objects of all kinds appeal to him. He likes to put them in his mouth and bite on them, probably to ease the discomfort of cutting teeth.

He is also learning another skill—walking. Seven days after his birth his first attempt happens as if by chance. While he is crawling he manages to straighten his forelegs and then his wobbly hind legs. And so he raises his body into a walking position.

Now he starts placing one foot and then the other forward. At first his left and right legs don't keep an even pace. This twists him up and causes him to sprawl flat on his face where he starts yelping *kyot kyot* in frustration.

Even after he gets his feet to working properly he still staggers, stumbling over pebbles and hummocks and hollows to take tumble after tumble. Finally, lying in a defeated heap, he screams shrilly, refusing to get up and try again. His screams, mixed with long, drawn-out shuddering sobs, arouse his mother. She gathers him up and comforts him and then decides to help him.

Placing herself a short distance from him, she crouches down, bottom up, face close to the ground. She looks at him intently and performs the rhythmic lip movement, in and out, in and out, which seems to say, "Come on! Come on! Try it."

The baby looks at her a while, then he too begins to move his lips, in and out, in and out. Like the smile of a human baby, the lip movements serve to relieve tension.

Finally the baby ventures again, straightening his legs and tottering toward her. His mother next walks ahead of him, then turns and waits for him to catch up. With every step his wobbly legs become stronger.

In the days that follow he finds the courage to practice by himself.

He totters along concentrating on keeping his balance, muttering *k.k.k.k.* When he stumbles over rocks he cries *e e e* in exasperation, his shoulders twitching in nervous spasms. He gets no reaction from his mother, but he expects none. He is really only talking to himself. In a minute he is up practicing again.

It takes only a week for his forelegs to become firm and straight. In another week, the hind legs also become so. By the time he has control of both hind and front legs he has developed the peculiar galloping gait characteristic of all Japanese monkey babies.

Now, at three weeks of age, he reaches a new phase of life, one of widening horizons and growing interests. Many dangers await him in a world that is daily enlarging around him, and he will not know by instinct what most of the dangers are. He must learn about them from his mother. Yet, many times, he will rebel at her teaching. The bond that still binds them closely will then begin to stretch a little.

# Life in the Troop

Through the lengthening spring days the life of the troop flows in patterns of wandering, foraging, resting—each segment lasting thirty minutes to an hour. The baby becomes more and more aware of this flow as his attention begins to turn outward. But all that he sees is still filtered through his mother's eyes, and his reactions are imitations of hers.

Life is rather well organized in the monkey troop. The leader occupies the center of power assisted by his two subleaders. He is also aided by the chief female, who helps keep order among the females. She is the mother of the first subleader and head of a powerful clique composed of her daughter and granddaughters.

The young mother belongs to another clique, one made up of her female family members. It is not as powerful as the chief female's clique, but it is growing stronger as the number of females born into it increases. Between these two cliques there is a constant struggle for power, which often results in bitter fighting. When that happens, the leader and subleaders charge in among the battling females with a bite here and a cuff there until the contestants scatter and order is restored.

Outside the cliques, and related to them only distantly, are several ordinary females who have no power. Usually they try to avoid quarrels by scuttling away whenever a fight breaks out, so they manage to live peaceful lives.

The troop's center—made up of leader, subleaders, and females—is surrounded by ordinary males. Their numbers are increased every year by male juveniles who at the age of five years drift from the center to the periphery.

Peripheral males have a ranking order too. It seems to be fixed in great part by the way they discharge their duties. These consist of scouting, helping round up laggards, and policing quarrels among their numbers. There is constant tension among the young males as they strive to raise their status. But no matter how hard they struggle, it is practically impossible for them to cross the strong but invisible barrier that keeps them from becoming a leader or subleader at the center of their troop.

Every now and then a discontented male in the periphery will relieve himself of frustrations by trying to cow the females, thus proving his superiority over them. One day, during a rest period, the third-ranking peripheral male comes upon the young mother as she lies stretched out resting with her baby.

Making gruff sounds, he flings himself on the ground before her in the grooming posture. When she does not respond, he snatches the baby out of her arms and tosses it aside. The baby lies there motionless, too terrified to make a squeak.

The mother leaps up, her mouth stretched wide open defensively, and starts shrieking, *Gya! Gya! Gya!*

She tries to retrieve her baby, but the male monkey blocks her way and bites her on the belly. Separated from her clique, the young mother feels powerless because her rank as an individual is lower than the male's. She does not dare fight back. But she turns her head over her shoulder, her eyes flashing a despairing signal to other members of her family. They start rushing to her aid. But they are too far away to be of any help.

The old leader is nearby, but he is not fond of the mother or her family, so he chooses to ignore the racket. The subleaders and the chief female make no move either.

Once more the mother reaches for her baby. Once more the male monkey snatches it up and tosses it aside. This time the baby, taking his cue from his mother, also screeches. His shrill shriek strikes home to the leader, for the cry of an endangered infant is never ignored by the monkeys.

The leader jumps to his feet, the hair on his head erect, his ears flattened, his tail raised. He protrudes his muzzle and stretches the corners of his mouth down, baring his long canine teeth in a threat grimace. Low but ferocious growls explode through his lips.

The subleaders, as if at a signal, spring to the leader's side. The phalanx advances, shoulder to shoulder, heads lowered and muzzles thrust forward, growling in unison, bobbing up and down and smacking the ground with their hands.

The females scatter. The young mother retrieves her screaming infant and rushes off. And the aggressive peripheral male finds himself in the midst of a furious avalanche of tawny fur. Lowering his once-proud tail, he tries to flee. But the leader and subleaders surround him, snapping and biting, their long-fanged teeth, large and daggerlike and kept sharp by frequent tooth grinding, drawing blood again and again.

Finally the peripheral male makes the only move left to him. Turning, he presents his rump to the leader, thereby acknowledging submission. The attack ceases. The leader places his hands on either side of the vanquished monkey's thighs and lifts himself up until he is resting his full weight on the other's buttocks. In this act he proclaims his authority.

Afterward, the defeated monkey gingerly grooms the leader in further acknowledgment of submission. Harmony is restored.

It takes a little while for the mother to comfort her baby. But he has learned a useful lesson, the necessary place of fear in his life. The lesson will be repeated often now that his curiosity is growing with every day. Suckling at his mother's breast no longer fills his whole world. Even while nursing he keeps looking around and reaching out his hands to grab at something new and interesting.

Sometimes he tries to slip away toward a strange object that has caught his attention, but his mother will never let him stray more than two feet from her. When he tries to go farther afield, she grabs him by an arm or a leg or his tail and hauls him back again. If her hands are occupied with grooming as he heads off, she just plants one of her feet firmly on his back, holding him in place despite his whining and squirming.

She still will not let him play alone with other infants. But sometimes she indulges him by carrying him over to an aunt who has a small infant of her own. Sitting opposite each other and holding their infants safely against their breasts, the mothers allow the babies to clasp hands and do some miniature boxing.

Meanwhile, as the baby exercises more and more, his limbs become stronger and firmer. He invents climbing, jumping, and swinging games that he plays by himself. He no longer has any difficulty walking.

By the time he is a month old he grows tired of clinging upside down to his mother's belly when she walks. Now that he is bigger, this position is very uncomfortable. His back scrapes over rocks and his head sometimes bangs against them. And the upside-down view of his mother's breasts, once so comforting to him, becomes a small stifling prison.

One day while his mother is on her haunches he scrambles onto her back. When she gets to her feet, he finds himself high astride her, his tiny thighs gripping her flanks, his body flat against hers, hands clutching her fur.

Entranced, he rides on her back when the troop sets out on the day's trek. From his high perch he looks about triumphantly, though somewhat shakily. Now the great world through which they forage unfolds before him in all its variety. Sunlight shines in slanting rays through the tall trees. Overhead the squirrels chatter, tossing down their leavings as they strip pine cones of their seeds. A brown hare dashes across an open space pursued by a red fox, silent as a shadow.

At first the baby sometimes finds himself slipping from his high perch. With a squawk of dismay he begins painfully pulling himself up again. Gradually, however, his muscles strengthen and he is able to keep his balance. Presently, holding on tightly, he can ride on his mother's back even when she climbs the highest trees.

Shortly afterward, his mother starts training him to walk with her, at first under her belly so that she can shelter him. Then, as he grows bigger, he ambles at her side. Now he has two ways to travel. On long or fast treks he rides on her back, while on short or slower journeys he walks.

Finally he masters another skill—sitting without support. He practices by leaning against his mother or clinging to something to prop himself up. If his grip loosens, he slouches back on his haunches. But as his hips grow steadier he is able to sit even on tree branches, freeing his hands for another activity about which he is becoming curious—eating.

Fascinated by the food his mother puts into her mouth, he bends over to smell each bunch of leaves or grass or flowers she picks. He likes to play with the bits of food that drop from her lips as she chews. Picking them up daintily between forefinger and thumb, he pops them into his mouth and then spits them out.

One day he pulls a leaf from a bush and puts it in his mouth as he has seen his mother doing. But he doesn't know what to do with it, so he pulls it out again—in and out, in and out, just playing. His mother gives him a lesson by grabbing a particularly choice leaf he is playing with and eating it herself. In this way she shows him that what he has been treating as a plaything has a useful purpose.

He is growing out of his baby talk, passing on to an enlarged vocabulary, although it is not yet that of the adult. Adult language floats all around him in a tapestry of sounds that change with the hour of the day.

In the early morning, before the monkeys set out on their trek, the air is filled with brisk calls. It is a different, more relaxed atmosphere when they gather at sunset at their campground. The last brief hour before the sun finally sets is spent in grooming by the adults and in frolicking by the older youngsters and juveniles. These gambol everywhere, scuffling, wrestling, and playing tag. The adults, far more particular about keeping their distance from one another, make up for a lack of closeness by noisy vocalizations.

Their vocabulary consists of more than thirty varieties of sound—soft *kus*, whistles, and warbles . . . squawks and squeaks . . . growls, squeals, and screeches. It is a rich tide of sound in which emotions from fear and sorrow to anger and pleasure and affection find expression.

As the sun rings down, the surge of voices dies away. One by one the monkeys climb to their nesting places in the trees. By twilight only the soft *ku ku ku* of questing voices floats here and there among the trees. Each monkey, tucked in the fork of a tree, is calling out for a last reassurance before night falls, that encroaching mysterious darkness which will soon engulf these creatures of the daytime.

Now the dusk begins to stir with its own life. Flying squirrels with their huge protuberant night eyes glide out of their trees on furry sails. In playful groups they bank and glide, snatching at one another on their way down with chittering *tjuk tjuk tjuks*.

Nightjars that have remained invisible all day float through the trees on broad gray-brown wings. Their monotonous song, *errrr*, is punctuated now and then by a shrill *shrui* and a fierce clapping of wings as a male performs his courting ritual.

As night deepens, the monkeys' last calls fade into silence. High in the fork of a tree, the baby, pressed against his mother's breast, suckles briefly at one of her nipples. Then, still holding it in his mouth for comfort, he dozes off. Sometimes, as dreams chase their way through his sleep, he babbles *k.k.k.k*. And sometimes, feeling an urgent need to urinate, he wakes suddenly, drawing slightly away from his mother so that he will not wet the soft furry nest.

Now the night enfolds him with its mystery—high cold stars or a moon that casts an eerie light. Somewhere out there the raccoon dog is roaming. The most primitive wild member of the Canidae, the canine family, he resembles a raccoon in appearance and coloring, though he is no relation. As he closes in on his kill his hunting cry, a meow followed by a growl and then a long, drawn-out whimper, floats through the night.

*Ho ho ho*, the intermittent hoot of the Ural owl, assaults the baby's ears. A chill wind ruffles his fur.

His mother, sleeping soundly, seems far removed from him. Only the loneliness is there, ballooning up until he trembles. His vague unease finds voice in thin sighing sounds that leak through his pursed puckered lips, *phili pili pili pii pii*. Whispering to himself, he turns and creeps back into his mother's breast and seeks her nipple again. He feels her arms tighten about him even as he falls asleep.

*Flying squirrels and nightjar*

$$5$$

# Forest Dangers

$S$ummer arrives quietly in the Alps. Spring flowers fade, and their places are taken by summer blooms. Oxeye daisies flutter white in the open spaces. In shady woodlands skunk cabbages raise their great white hoods, sheathing greenish spikes studded with flowerets. The wild-cherry trees have scattered their white blossoms, and the small knobby fruits will soon be ripening.

In the forest, life takes on a lively tempo. Kingfishers piping *pi pi pi pi* skim the swollen streams in search of very small fish to feed their young, hidden in nests scooped out of the steep banks. Sometimes at night the shriek of a female badger ready to mate rips through the darkness like the cry of an injured child. Daytime is disturbed by the anguished squeals of a female hare being buffeted into the mating position by a male.

The monkeys' birthing season, which lasts from mid-April to mid-July, is over for this troop now that the last pregnant female has given birth. One of this season's babies was stillborn, but for a whole month its mother would not accept the fact. She licked it and cuddled it and took it with her on foraging trips, clasping it to her breast.

During rest periods she would sit alone day after day because others would not allow her near them as long as she carried the corpse. Her shoulders drooping, she would cry softly, *ku ku ku ku*, now and then stopping to groom the dead baby. And all the while her eyes would keep probing into the far distance, as though in a hopeless search for something lost forever.

As the days went by and the tiny body shriveled, she stopped cradling it and began dragging it over the ground as if it were an inanimate object. Then, one day in mid-June, when the monsoon rains began to dump their moisture on the Alps, she discarded the body altogether.

The rains, coming from the tropical south, bring hot, humid days. This is the monkeys' molting season, which begins in May and lasts through July. First to molt are the males and those females who did not bear young this year. Pregnant females do not begin their molts until their offspring are born. Now the young mother looks like a moth-eaten old coat from which swatches of long pale gray hair have been stripped. Soon she will be clothed in a shorter, buff-colored summer coat.

Changes are being wrought in the baby monkey, too. His face begins to grow longer. Black pigments settle in the skin of his hands and feet, giving them a darker appearance. Canine and premolar teeth erupt along with the rest of his milk teeth.

By mid-August, when the monsoons are over and hot, sun-drenched days have taken their place, the baby sheds his downy baby hair, first from his buttocks, then from his legs and arms, and finally from his head, back, and sides. He sprouts a new coat of buff-colored adult hair.

Though he is still suckling, he starts foraging with his mother, learning how to stuff his tiny cheek pouches so that he can eat at leisure. He is inquisitive and daring enough to poke anything within sight into his mouth, and this presents a new danger. Some of the plants and toadstools are poisonous, and he does not know the difference by instinct. He must be taught what is good and what is not.

His mother is his teacher. When she sees him reaching for something that is not eaten by the troop, she grabs his arms and jerks him away, scolding him with sharp yaps. Sometimes the plant he is reaching for is harmless and may even be the staple food of monkeys in other places. But the baby must not eat it if it is the tradition of this troop to shun it.

One day he may manage to elude his mother and sample a strange plant or tempting fruit. Then he will either become violently ill—perhaps even die—or he will be able to introduce a different and valuable foodstuff to his troop. It is through their young that the monkeys acquire new habits. The adults are too conservative ever to change their ways.

Foraging is a delightful occupation to the monkeys, and they greet each day's venture with joyful anticipation. The tide of gaiety that sweeps through the troop when they waken in the morning infects the little monkey. Dancing and gamboling about his mother he chants *k.k.k.k.* and *hoii hoii hoii . . . pii.*

The trails are now summer ones, leading over the cooler northern slopes of the mountains and down into the deep shady ravines. Here the monkeys feed on the tender shoots of young bamboo and stalks of spikenard. They also catch and devour insects of all kinds—grasshoppers, cicadas, beetles—and even crabs, which they scoop out of the dwindling streams.

The baby watches his mother deftly snatch up grasshoppers poised for flight, or cicadas newly hatching on tree trunks. One day he grabs a

*The kingfisher*

grasshopper himself, holding it by the head and watching its wings churn and its legs kick—a plaything. His mother snatches it away and eats it.

As the dry season progresses and grasses and flowers die and leaves grow tough and unedible, the monkeys' food supply dwindles. They have to forage for much longer periods to get enough to eat. The heat saps their vitality, especially during the early afternoon. They spend as much as two hours at a time in the cool shadow of the giant cryptomerias, the oaks and beeches.

Adapted to a temperate zone, they suffer more in hot weather than do most other monkeys because they neither pant nor sweat heavily enough to maintain their body equilibrium. To make up for this they must indulge in more frequent and longer groomings. In grooming, the mat of fur is loosened and separated, letting the air penetrate to the bare skin.

The heat does not affect the baby much. While his mother dozes he is as active as ever. He makes a game of teasing her, stealing up and tickling her face with his fingers. Annoyed, she pushes him away, threatening him with a slight opening of her mouth. Off he dashes, only to return to his teasing the minute her eyes close.

A strange relationship has sprung up between the two. Both are torn by a desire to keep the close bonds that have united them since the baby's birth. At the same time, both strain to break free of each other.

Sometimes the mother allows the baby to play with other children his size as long as he stays within a fifteen-foot radius of her. Sometimes she even permits his two-year-old cousin to carry him around and amuse herself with him. At other times his mother keeps him anchored to her side with an iron grip on tail or leg, while he struggles, howling to be free.

When he wants to lie in her embrace and suckle, she often shoves him away roughly and walks off. This sends him into a temper tantrum.

Screeching *Gyaa! Gyaa! Gyaa!* he chases after her. Finally he throws himself on his stomach, back arched, and sobs in long, gasping shudders, his shoulders jerking at every breath. At last his mother relents. She picks him up and lets him nurse, but, worn out by his tantrum, he can only continue to sob softly until he falls asleep.

As he grows a little older and wiser, he learns not to exhaust his strength in real tantrums. Instead he simulates them, rolling himself into a ball and screaming without any real emotion. Every now and then he raises his face to look at his mother and see if she is moved by his behavior.

When he realizes she is not, he starts playing by himself again, or he jumps up and buries his face in his mother's breast and hugs her, crooning affectionately and coaxingly, *ngu ngu ngu ngu ngu.*

His mother often ignores these endearments too, so that little by little he learns that no matter how much he pouts or coaxes, he is losing his position in the center of his mother's world. But she never ignores him when he gives his sharp shrill shriek, *ki ki ki ki.* Even though she does not know what has frightened him, she rushes to him and gathers him to her protectively, then looks around to find the cause. It may be anything, for the young monkey's sense of fear and caution has been growing rapidly.

Now he will cry out at a harmless red fox, a tawny badger sunning itself at the mouth of its den, even a small wood mouse scampering across his path. His fear subsides only when he sees that his mother herself shows no concern.

But some forest creatures should inspire fear in the monkeys. One day the troop surprises a female black bear, its broad chest marked with a white crescent. With two cubs in tow, the bear is foraging for food—insects, ripening berries, honey, small water creatures. As a rule the Asian black bear, which is one of the smaller bears in the world, minds its

own business, but it can be dangerous, especially when it has young to protect. Shoving her cubs behind her, the bear turns at bay.

The nearest male to spot the bear rises on his hind legs and cries *Kuan! Kuan! Kuan!* The warning is relayed through the forest, as mothers and children scramble up the trees. The old leader hurries to the place where the alarm was first given. Surrounded by subleaders and other males, he leads a chorus of growls. Intimidated, the bear turns and lumbers off, herding her cubs before her.

Few forest predators can harm the adult monkeys with their quick eyesight, nimble responses, and well-organized group defenses. It is the infant monkeys that face the greatest danger. Until they become juveniles able to care for themselves, their safety depends largely on the care their mothers show them. And not all mothers are good ones.

Some mothers begin to neglect their babies within a few days of their birth, leaving them sprawled on the bare ground while they feed themselves. Others try to push their week-old babies from their breasts to their backs. Some are always snapping and biting at their young, cuffing them whenever they approach. And when their youngsters grow older, they are allowed to wander far afield when foraging because their mothers do not want to bother with them. Often the daughters of such mothers grow up to treat their own youngsters in the same way. It is these neglected young monkeys that make up most of the fatalities in the troop.

One day, a seven-month-old female is unable to scale a low cliff on the troop's trek and is left behind by her careless mother. Screaming unheeded, the baby sprawls on the ground. From some hidden sentry post, a golden eagle spots the shrieking infant and soars upward, crying harshly, *eak eak eak*. Its seven-foot wingspan casts an ominous shadow on the land below.

What a flurry there is in the forest! Squirrels dash for their holes.

*Forest Dangers* : *211*

*The menacing black bear*

Hares race for cover. Monkey mothers summon their young, even the two-year-olds. Sheltering them in their arms, they scream *Kuan! Kuan! Kuan!* and *Ke! Ke! Ke!* at the soaring eagle.

Ignoring the cries, the eagle swoops down on the prostrate baby monkey. Its wings create a rush of air, its huge talons sink into the baby's soft skull. Agonized screams ring through the forest. But before the eagle can get airborne, the old leader and several other males come plunging over the cliff and lunge at the great bird with gnashing teeth and flailing hands. The eagle lets go of the baby and flaps off, barely escaping being mauled by the monkeys.

The little monkey seems only half alive. The flesh in the soft crown of her head has been torn open, revealing a gaping wound surrounded by crushed bone. She sits with her fists doubled up on her breast, an expression of blinding pain on her small face.

Suddenly one of the adult males hurries over to her. Tenderly he gathers the injured baby up in one arm and carries her carefully to that night's campsite. The next day, while the others go on their daily trek, he stays with the injured infant, taking her with him while he forages, cradling her for hours in his arms.

At first the pain is so intense that the little creature cannot move or even bend over to forage. She can bear to eat very little. Hour after hour she sits motionless with her doubled-up fists still pressed to her chest, the anguish plain on her contorted face.

But, as the days pass, the gaping wound begins gradually to heal, leaving a great bald scar on the baby's head. The pain is gone, however. She is able to move about again. The male lets her scramble atop his back, and together they hurry to join the rest of the troop.

From then on, the male adopts the baby as his own. The mother scarcely cares. Besides she knows she could never get her daughter away from the male even if she tried. The little female rides her foster father's back when they forage and sleeps in his arms at rest periods. Sometimes he grooms her, or the two engage in a gentle wrestling game, playfully boxing and biting at each other.

Even after the little monkey grows up, her relationship with this male will be that of father and daughter. They will never mate. Japanese monkey parents do not mate with their offspring, and this rule also applies to most of the adopted monkeys.

# 6

# The Mating Season

September is the month of typhoons. They usually strike the southern parts of Japan; only their fringes sweep over the Alps. But they bring high winds and torrential downpours, which account for much of the 130 inches of rain that fall yearly here.

Streams turn into raging floods. Loosened rocks and mud tumble down steep slopes, carrying everything before them. Danger lurks on all sides for the creatures of the wild.

Though the coats of the monkeys shed much of the rainwater, the little animals are miserable. During the heaviest of the downpours, foraging is difficult. The monkeys must spend their time sheltering in the trees or under overhanging rocks.

But, finally, the typhoon season passes and the glory of October arrives with sparkling days of sunlight. Night frosts transform the green needles of the larch to pure gold, matching the golden clouds of birch and beech. Graceful maple trees flame scarlet against the spires of green firs.

And brightly colored leaves, loosed by the breezes, drift through the air like confetti.

This is the harvest season, the richest of the year. Hazel nuts, chestnuts, acorns, wild grapes, and the varicolored berries of autumn are all ripening. No matter what trail the monkeys take now, it will lead them to an autumnal bounty. Thus they wander widely throughout their territory.

The baby monkey, though still suckling, eats ravenously. At five months of age he is a chubby, bouncing ball of buff fur weighing almost four pounds. His mother is quickly regaining the weight she lost after bearing him. Her fur is soft and glossy. There is a general appearance of well-being throughout the troop.

Overhead the squirrels are stuffing themselves too, as well as gathering nuts and pine-cone seeds by the thousands. Carrying them in their cheeks, they scamper off to store them in tree holes and discarded birds' nests with an eye to the coming winter.

The small red-tailed shrike is also preparing its winter larder. On thorns or sharp tree twigs it impales a part of every day's catch of small birds, mice, lizards, toads, and worms. There the tiny dried carcasses will hang, insurance against the approaching lean season.

The starlings are gathering for their fall migration southward. In flocks of thousands they wheel and bank, ascend and descend, in perfect order against the sunset sky until, as twilight falls, they settle to earth in a dark cloud. Soon they will all be gone.

By mid-October a change comes over the adult monkeys and older juveniles. The faces and genitals of both males and females flush from pink to a deep scarlet, causing the twin gray pads on their rumps to stand out plainly. Hormonal changes are taking place in their bodies, rousing the mating instinct.

The females lose their appetite and become nervous and high

strung. The males strut about, their tails raised like flags.

The urge to mate draws in two male visitors, strangers who suddenly appear on the outskirts of the troop. They arrive from different directions. Several years ago, both deserted their own troops to live a solitary life in the forest. One is a husky mature adult some thirteen years old. The other is old and ugly, with tattered unkempt hair and filthy teeth. He creeps about in a senile way, his hands trembling. Both these mismatched males will stay with the troop throughout the mating season. During that time they will be ignored by the other males. Afterwards they will either leave again or try to make a place for themselves here. This will be difficult because, once the mating season is over, the troop will start showing animosity toward them.

The season begins in earnest when the old leader climbs a tree and shakes it furiously, roaring *Ga! Ga! Ga! Ga!* Though many monkeys shake trees, only the high-ranking ones roar as they do it.

Now, all day long, during rest periods and foragings, the baby monkey sees the old normal routines turned upside down. Even his own mother, on whom he has always relied, is swept up by new urges.

Noisy clamor disrupts the day—squawks, squeals, and yaps vie with screeches and screams. Trees suddenly burst into violent commotion as they are shaken by fierce-eyed monkeys. From time to time, roars accompany the tree shakings, the sign that a high-ranker is doing the shaking.

Sometimes the warning cry, *Kuan! Kuan! Kuan!* rings through the air. At any other season it would mobilize the troop. But now it is recognized as the female's mating call and is ignored. Underlying all the commotion is a soft intermittent murmuring of long, low *kuuuus*.

*Kuuuu. Kuuuu. Kuuuu.* It rises and falls. Now and then the baby hears his mother crooning also, but her *kuuus* are not directed at him. She seems to have forgotten him as she wanders about.

*The Ural owl*

Everywhere, strange antics are to be seen. Here a stiff-legged male prances toward a female, gritting his teeth and smacking his lips. Another is kicking up his hind legs in a stiff jumping-jack pose.

Couples are grooming each other, either on the ground or in the trees. But now there are few female-to-female groomings and no male-to-male ones. Females are grooming males, and males are grooming females. Here and there, mountings are going on. Most of these are not to show the dominance of one male over another. They are true matings of male and female.

In this topsy-turvy season the ranking orders of females and males seem almost to disappear. High-ranking females sometimes mate with ordinary males. High-ranking males mate with ordinary females. It is all as fancy chooses.

Females may even reject the old leader's overtures simply by not standing still when he indicates his desire to mate by pushing lightly on their backs or hips. More than once, the baby monkey sees females turn away from the old leader's courtship proposals to pair with others of no rank at all. In this melee the solitaries, who would get short shrift from the troop at other times of the year, are ignored for the most part as they join the wild mating orgy.

Few unions last very long. Males and females change mates frequently. A mating may take a day or may continue for weeks before the pair splits up to look for different partners.

Now a male, smacking his lips and grinding his teeth, starts prancing stiff-legged toward the baby's mother. He is a male of high rank, and so he makes his proposal flamboyantly. When he comes up to her, he suddenly leaps right over her. He turns and repeats his leapfrog antics two or three times.

Far from being cowed by this behavior, the young mother resents it. She looks over her shoulder at the offender with a wide grimace—half fear and half anger. When he grabs her forearm, she squawks at him. The baby mimics her grimace and her squawk.

Finally she is able to jerk away and races to a nearby tree, scaling it with the baby at her heels. Not to be put off, the male comes shimmying up after them. Higher and higher climb mother and son, until they can go no farther.

Higher and higher the male clambers after them. But the thin branches cannot support his weight. As he stands on one it cracks and breaks and he tumbles head over heels, catching himself on a lower limb. Chastened, he slithers down the tree and goes away.

Soon the mother and baby come down too. Just as they reach the ground the mother's attention is caught by an adult male's movements. She utters a soft *kuuu kuuu kuuu* in his direction. But the male only gets up and walks away.

The mother, with the baby tagging after her, follows at a distance. Gradually she draws closer, giving occasional *kuuus*. When she is only some twenty feet away, she takes a roundabout path that brings her to his side. She begins grooming him, but after a few minutes he walks off.

She shrieks as he goes and starts to follow him, then gives up. She has been rejected just as she rejected the high-ranking male. She sits down disconsolately, murmuring *ku ku ku* as though to herself. Then she resumes her wandering.

The baby tags along growing tired and bored. He begins to lag farther and farther behind. Finally, when he can endure his sense of isolation no longer, he utters a soft, pleading, *ku ku*.

It is enough to recall his mother. As though suddenly remembering him, she turns around and quickly returns. She begins to groom him. The old feeling of security comes back as she takes him in her arms and lets him suckle.

Sometime around mid-afternoon a six-year-old male, newly arrived at adulthood, accepts the young mother's proposals. They grind and chatter their teeth and smack their lips at each other. Then he begins to groom her. Presently she grooms him. They converse as they groom, *kuuu kuuu kuuu*.

How left out the baby feels! He stretches himself on the ground before his mother in the grooming posture. Always before when he has done this she has been quick to respond. Now she ignores him. He begins coaxing her in soft nasal whines, *ngg ngg ngaar*. She ignores this too. He turns over in another posture and again coaxes. But she has no eye for him. All her attention is focused on the male.

Suddenly she stops grooming and with one hand slaps the ground smartly. It is her signal that she is ready to mate. The male places his hand on her back and pushes her lightly forward from the waist. It is his signal that he too is ready.

The young mother stands up on all fours, and the male puts his hands on her hips and wraps his hind legs around hers just above the knees. She has to stand stiff-legged, supporting his full weight as the mating takes place. It lasts sixteen minutes but will be only the first of several mountings. There may be as many as nineteen before the mating is over.

Now the male dismounts, and the two sit together for a second or two uttering several *kuuu kuuus*. Then the second mounting takes place, followed by a third, a fourth, and a fifth, with shorter rest periods between. The mating is completed on the sixth mounting. A shudder passes through the male. He grunts *uh uh uh*. Making soft cackling sounds, the female turns her head to look at him as she reaches backward with one hand to grasp his beard.

Then the male dismounts. Both sit side by side in a peaceful lethargy. They are suddenly roused from it by a series of roars, *Ga! Ga! Ga! Ga!* that break out behind them. They look around to see a male approaching in a stylized trot.

The newcomer ranks higher than the young adult with whom the mother has been mating. Ordinarily he would attack the low-ranker for his temerity. But now his attention is focused solely on the young mother.

One hand reaches out to grab her and maul her, not as a punishment but as a means of attracting attention to himself. It is actually a mating proposal. The young mother wants none of it. She jumps to her feet and flees, but the roaring male is in close pursuit. The one with whom she has just been mating starts to follow, then thinks better of it and walks off in a different direction.

As the young mother disappears into the forest undergrowth her abandoned baby tries to trail after her, but he is unable to keep up with the mad chase. He wanders about, disorganized, desperately searching. Fear seeps through him. Never has he been so utterly separated from his mother. Always before he has known where she was.

He calls *hui hui hui* in a quavering voice as he looks for her. An old female, touched by his distress, tries to embrace him, but he escapes and continues on his way, uttering that plaintive cry over and over, *hui hui hui.*

His young cousin hears his cries and also hurries to comfort him, but he ignores her too. On and on he searches.

Shadows grow long. Chill night breezes shiver through the forest. The baby's fear grows until he is in the grip of a terror dark and terrible. He starts screeching in panic.

Then suddenly he hears that familiar voice answering him, *ku ku ku ku.* His screeches stop and his answering voice takes on a musical lilt as he calls again, *hui hui hui.*

There is his mother. Whistlings of relief come with every breath as he races to her and she catches him up in a tight embrace. That night, as on every night in the past, he sleeps safe in her arms, against her furry breast, one of her nipples in his mouth.

# 7

# Cruel Winter

As the mating season progresses, the baby becomes used to these separations from his mother. As long as he is among other members of the troop, or with his young cousin, he no longer feels the uneasiness that overpowers him when he is alone.

His mother has embarked on a new courtship with a fresh partner. She will have several mates before the season ends. Some of the relationships will be very brief; others will last several days, perhaps weeks. But she will not neglect the baby. When he calls, she will come and suckle him. At night she will sleep with him held in her arms in a warm embrace.

Meanwhile, November approaches. With the growing chill the monkeys begin shedding their light summer coats for the thick long-haired grayish ones that will see them through the winter.

In the early days of the month the monkeys are still able to find a good supply of food. Some of it comes from the farmers' apple orchards.

Perhaps the insecticide sprayed on the apples is distasteful to the monkeys. But, several years before, a juvenile accidentally dropped an apple in a nearby pond, and when he fished it out and ate it, he found the apple tasted better. From then on, he started washing his apples. Many of the younger monkeys began to imitate him. Now most of them squat over ponds or streams scrubbing the fruit thoroughly, smelling it, washing it, smelling it again and again until, finally satisfied, they munch on it.

The apples do not last long, for the farmers soon harvest the fruit. Autumnal gales tear the last leaves and nuts and fruits from the deciduous trees. The poplar, birch, beech, and maple that flamed so brilliantly are now black skeletons. It will be five months before they will again be clothed in fine new greenery.

Everywhere, bushes and woody vines die back. Cicadas, frogs, and crickets, which provided food in the summer, disappear. But the monkeys can still glean enough sustenance from the stalks and leaves of herbs, the pith of creeping plants, and the seeds of drying grass.

As the month progresses the days grow shorter and more chill. Dank white fogs appear, and sodden rains sometimes turn to sleet.

The Asian black bear, fattened on the autumn's plenty, begins to feel the chill in its blood and seeks out a cave for its winter hibernation. The plump sleek badger, too, curls up in its den to sink into a deep sleep from which, however, the least disturbance will wake it instantly.

In the early days of December the first snow drifts down from gray bellied clouds to dust the trees with powder. From then on, snows are more frequent. Soon the whole world is shawled in a wintry cowl.

The baby monkey, warm in his thick wooly coat, finds the glistening white landscape a strange and wonderful place. The feathery powder is soft under his feet. He kicks at it tentatively, and snow sprays outward. Soon, like the other monkeys, he is jumping gleefully in it, or sliding merrily down steep embankments. He watches curiously while his

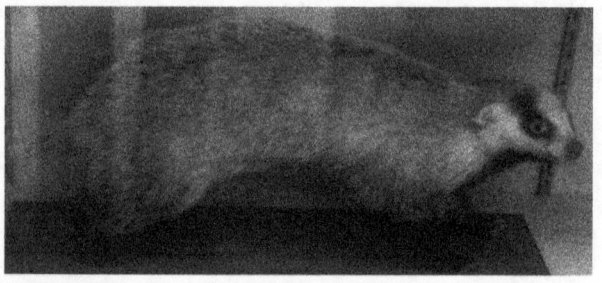

cousin makes a snowball and walks around, carrying it under one arm. Then he tries to imitate her. Many other monkeys are playing with snowballs in the same way.

Snow is collecting in the high ravines, but on the lower mountain slopes the monkeys can still graze on the tips of plants and shoots of vines poking through the thin veneer of whiteness. There are also clusters of scarlet winter berries that ripen in late fall and early winter. But before January arrives the last berries wither, drop, and fall and are buried in snow.

Now winter is upon the land in earnest. Snow six feet deep blankets steep slopes and valleys. The frozen surfaces of smaller streams and ponds are heaped with it.

In their moss-lined nests deep in hollow trees the squirrels lie snug, their bushy tails wrapped around one another for warmth. Only on sunny days will they venture out to raid their summer larders.

The marten, his brown summer coat exchanged for one of pale golden fur, and the red fox stalk the small creatures that are still afoot these wintry days. The hare is their principal prey, but he has become elusive, camouflaged perfectly in a coat of snowy white.

For the hare, as for all herbivores, life is grim. His food now consists chiefly of twigs and bark, although here and there he is able to find some lichen on exposed rocks. The shaggy-haired serow, the Japanese antelope-goat, also makes his meals off dried branches and twigs. This becomes the sole food of the monkeys too.

Scrambling through the trees, each monkey selects its own branch. It grasps the branch in both hands while standing upright on a lower branch. With its sharp strong teeth it strips off the bark lengthwise, leaving it dangling. Then it nibbles on the thin layer underneath, that layer from which next year's growth would have formed. Many trees bear deep, ugly scars made by the monkeys in their quest for food.

*Cruel Winter* : *225*

*The shaggy-haired serow*

Sometimes, when the bark on larger branches is so hard it cannot be stripped, the strongest adult monkeys are able to gnaw a whole piece off. When this happens, the fortunate monkey holds an end in either hand, bites the piece in two, then sits down to eat both halves.

The baby cannot peel the bark from the larger branches, and so he nibbles on twigs. Along with the other young monkeys he also learns to chew on the trees' winter buds, which contain the shoots of next year's growth tightly coiled in hard knots.

The leader still sets the pattern of the day's foraging activities. But actually all movements are now at the whim of nature. Storms and deep snowdrifts prevent the long treks of autumn. At best the troop travels for only short distances and then mainly through the trees, swinging from branch to branch, babies clinging to their mothers' bellies. When open spaces have to be crossed, the troop marches in single file, led by the highest-ranking ordinary male. Head sunk in snow he plows out the course to the next stretch of forest.

One day, the monkeys find a small respite from the cold when they come upon a dying campfire left by mountain climbers. Eagerly they crowd around it warming themselves. Entranced by a glowing ember, the baby dashes in to seize it. But as he touches it he lets out a shrill squawk of pain and backs away into his mother's arms, sucking his burned finger. He has learned another lesson.

Not all the days are stormy or gloomy. Sometimes a sparkling sun comes out, glittering on the cold snowfields. Then the monkeys enjoy quiet grooming parties in the leafless trees. Their mating season, which lasts into January, has long since peaked, taking with it the exuberance of fall. Only the juveniles seem possessed of an inexhaustible energy. They scuffle and squawk through the upper branches, stopping to feed now and then on the winter buds.

The infant who was injured by the eagle seldom takes part in these games. Though she has recovered from her wound, she has never regained her old vitality. Most of the time she is content to creep into the arms of her adoptive parent and lie there quietly.

In late afternoon, as the sun westers and the air grows more and more chill, the monkeys begin to feed again, the activity warming them. While they eat, the sun sinks lower behind the mountain ridges to the west, sending a lengthening shadow up the opposite slopes.

The monkeys move just ahead of the rising shadow, feeding until dusk settles over everything. Then they take to their nightly perches in the trees. In the summer the tall, thickly leafed cryptomerias were their favorite sleeping places, and the umbrellas their branches form still give the best shelter from the wind. But the monkeys avoid these trees when their crowns are weighted with snow, for sometimes the snow will dislodge during the night and deluge them with cold wetness. Instead, they choose the naked branches of the deciduous trees. There, forgetting their aversion to close bodily contact, they huddle together for warmth.

And still the temperature keeps falling, sometimes dropping below -17.5 Centigrade. As the world enters the very core of winter the leader directs the troop to a rocky ledge that thrusts out from the steep banks of a frozen stream. Above the ledge, which generations of monkeys have used, an overhanging rock shelf shields it from the worst of the weather. Here the monkeys make their base, going out to forage during the day and returning at night, or in fierce storms, to shelter against the far wall.

But one late afternoon while they are out foraging, a violent blizzard swoops suddenly down on them. In that blinding fury of gale and whirling snow it is impossible to keep up the trek. So they stop where they are, crouching in upon themselves to preserve as much warmth as possible.

As the storm grows fiercer, the individual monkeys begin to clump together. Finally the whole troop forms a tight-knit lumping that takes advantage of the collective heat their bodies generate. Here they pass the bitter night, hearing like an echo of their own distress the mournful howling of young foxes lost in the storm.

Early the next morning, the blizzard breaks and the sun shines out. The monkeys unscramble themselves and get to their feet. Only then does the foster male realize that the little adopted daughter he has been clasping in his arms all night does not stir. She has succumbed to pneumonia, but he cannot understand that she is dead. Clutching her lifeless body tightly, he croons to her. When she does not respond, he shakes her, but her head only lolls on her neck.

At last he stumbles to his feet and tosses the little corpse aside. Staggering through the deep snowdrifts after the troop he feels a strange emptiness where once the little female rode on his back or belly. It is as though an invisible bond had snapped, releasing him from the troop. Taking a different course, he trudges off by bimself and is soon lost in the white world. From now on, he will live a solitary existence.

Winter cannot last forever. As February gives way to March, the weather starts surreptitiously to warm. Little by little the snow is melting, revealing as it does a poor store of treasures that the monkeys hastily retrieve. There are snapped-off branches to be nibbled, winged seeds from the maple, birch bark and fallen leaves, old acorns and various berries buried the winter long.

As March gives way to April and the melting of the snow becomes more dramatic, great danger threatens the monkeys, for this is the season of avalanches. One morning, while the adults are grooming and the young are playing in the treetops, the leader suddenly roars a strident alarm, *Kuan! Kuan! Kuan!*

Over his shrieks comes a great roar, which increases with every second. A huge slab of snow at the top of the steep slope on which the monkeys are feeding has broken loose. Down it plunges, gaining momentum as it comes.

The adult monkeys and most of the juveniles scatter. But the baby, feeling secure on his high perch, ignores the warning. His mother looks around for him and sees him high in a tree in the path of the avalanche. She calls him frantically, *Gya! Gya! Gya!*

Trained to obey her squawks of danger, the baby scrambles down. When he is halfway to the ground his mother reaches up and grabs him and races for a rocky projection. Just in time. The furry white tide flings dislodged rocks before it sweeps down the slope past the two cowering monkeys. A biting spray showers them.

The baby's cousin is not so fortunate. Swinging in another tree farther down, she also has ignored the leader's warning. As the avalanche howls upon her, the tree to which she clings shudders violently. Then it snaps in two like a matchstick. The little female is tossed into the maelstrom, squawking in terror and pain. Her mother rushes to save her but is carried off too, her shrieks joining with those of her daughter.

The leader and several males, tails up, rush down the slope in chase, threatening and growling. But the white tide outdistances them, roaring into the deep ravine below. And the cries of mother and daughter are still.

Slowly, defeated, the monkeys turn back, tails lowered, to rejoin the troop.

# 8

# Growing Up

For a while, the little monkey will not accept the loss of his cousin. He wanders around looking for her and calling her. When she does not come, he sits crooning in a melancholy voice. But he cannot remain lonely too long.

April is well on its way. Everywhere, snows are giving way to spring rains. Streams freed of their ice rush madly down the steep slopes with a fierce grinding of boulders and pebbles. The winter buds are swelling and softening, and the monkeys feed on them ravenously.

On the slope cleared by the avalanche, the white and purple flowers of the coltsfoot are springing up. The monkeys come to graze on them.

All the world is quivering into fresh life as the snows recede. Grasses and herbs renew themselves. Bushes and vines put out fresh leaves and tendrils. The larger trees begin to unfurl their first tentative leaves. Spring flowers burst into bloom.

Again the troop starts to follow the greening trails of spring. On one of these treks the baby monkey squawks in terror as a shrieking roar, *Ga! Ga! Ga! Ga!* sounds from overhead, where a big tree is violently shaking. There stands the younger of the two solitary males who joined the troop during the fall mating. With the first thawing, he left to go his own way. Now standing tall, erect, face contorted in a fierce grimace, he challenges the whole troop.

Several of the younger males respond with threatening growls, but the leader, secure in the strength of his troop, ignores the roars of the solitary, so the other monkeys pass quietly on. At the rear of the column trudges the other solitary. He has managed to survive the winter, though he has become more gaunt and bedraggled than ever.

Unwilling to return to the lonely hermit's life he had been living for so long, he tags along after the troop, though at the expense of his pride. Now he must occupy the humblest position—even lower than the newly matured six-year-olds.

In this position he must submit to numerous dominance mountings by his many superiors and bear the jealous attacks of aggressive young males who occasionally band together to drive him away. He never goes far and presently returns, head bent, tail drooping, drawing as little attention to himself as possible. He knows that only by showing this submissive attitude will he be allowed to stay.

Winter has already cost the lives of three troop members. But the long hardships and meager diet will exact a further toll. Several pregnant females lose their babies before the birthing season begins.

Seven other females give birth successfully. But one of the newborn babies is seriously deformed. His stubby legs and arms end in mere nubs. His mother has given birth to two other children, both of them with minor deformities. On the hand of one three fingers are fused together, while on the hand and foot of the other there are two fused fingers and two fused toes.

Are these signs of poor heredity or are they the result of herbicides used to spray mountain fields and orchards, the poisonous mists drifting into the nearby forests? The baby's mother, of course, is not troubled by such questions. She does not seem even to be aware of her child's terrible handicap. Unable to walk or gather food, or to cling to its mother when she is trekking, it is doomed to an early death—in the heavy snows and blizzards of the coming winter, if not before.

This spring, the mother of the baby monkey, now a yearling, does not bear a child. Usually Japanese monkeys have children only every other year and sometimes every three years. This enables each mother to wean her child according to choice. Some mothers are more possessive than others and want to hold on to their children as long as possible. But the yearling's mother is quite ready to give her child his independence. Rarely now does she allow him to nurse.

Once, when he comes begging, she cuffs him away. One day she emphasizes her rebuff by grabbing him and giving him a hard shake and a sharp bite. Now he really howls. But in the midst of his howl he sees another yearling nearby. His screams stop instantly. With so many things to absorb his interest, these rejections by his mother are quickly forgiven and forgotten.

Leaving his mother, he prances toward the other small monkey, uttering soft, friendly *kus*. Soon the two yearlings are engaged in a tussling match. Presently other yearlings come to join them in their play. It is the beginning of a small yearling troop whose members will soon become inseparable. The acknowledged leaders are the two new friends who founded the troop and now form the nucleus of it.

As the year slips into summer the infant troop, made up of males and females, sticks together. They have by this time all been well trained in what to eat and are able to forage for themselves. They still accompany their mothers during treks and now and then join them for a grooming.

*Growing Up* : *233*

*The unnoticed badger*

Though they no longer nurse, they sometimes still like to be held by their mothers and play with their nipples.

But, most of the time, they enjoy a riot of fun with one another as the spring slips into golden summer. They scuffle and wrestle together, giving playful nips. They swing by their heels from branches, forming living chains by grasping hands. They play tag and mount each other as they have seen the adults doing. They shove one another over rocks, tumbling down in graceful falls only to clamber back up again to play the same game over and over.

They no longer see everything through their mother's eyes but now depend on their own observations. And they try to mimic all that the adults do. They copy the quarrels that break out in the troop as males or female cliques strive to raise their social positions.

The yearlings' quarrels are only play scufflings. But sometimes they become so boisterous that the yearling females retreat to their mothers, much preferring a quiet session of grooming. As so often happens in play, one or another of the juveniles oversteps the bounds and launches a real attack.

The attacked yearling, unable to withstand the fierce assault, may run squawking to his mother, pursued by the jubilant victor shrilling *ga ga ga*. He is brought up short, however, when his victim's mother gives him a quick cuff and a threatening snarl.

Now it is the cuffed yearling's turn to start squawking. This in turn brings *his* mother, and the fight is on. *Wham! Bang! Squawk! Squeak! Gya! Gya! GaGaGa!* The two mothers fight and snarl it out while the yearlings watch.

Eventually one mother wins over the other, with perhaps some help from a friendly male. When this happens, the defeated mother and yearling have to flee. Such squabbles help determine the pecking order of the young monkeys. The little yearling comes to realize that the mothers of some of his friends are stronger than his own mother. He submits to the bullying of these yearlings, but he in turn bullies others whose mothers are weaker than his. The ranking order the yearlings establish among themselves is not firmly fixed, however, and may shift from day to day.

Just as the yearling learns not to challenge small companions who can call in more powerful help than he, he also discovers that this same rule applies in the grown-up world. Over and over his mother has tried to teach him never to cross the path of the leading males. And often enough, lured by some cluster of tempting berries, he has ignored her warnings. Then she has had to rush in with sharp, shrill yaps to rescue him just in time.

But one day, away from her close supervision, he scampers directly in front of the old leader himself. To the yearling's shocked amazement the leader lifts him up in one powerful hand and tosses him out of the way. He lands with a squawk of pain and rushes to his mother, but she has seen it all and also gives him a cuff to remind him of a lesson badly learned.

The yearlings do not just mimic the adults' quarrels, they also copy their games. One of the favorites played by males and females, young and old alike, is the tree-shaking sport. Sometimes the males line up for it. Beginning with the lowest-ranker and moving up to the highest, each climbs the tree and gives the branch a shaking. The higher the rank, the greater the shaking. Finally the old leader himself, though he may not have been in the game to begin with, comes rushing over, unable to resist climbing the tree and giving the final shaking with the fiercest roar.

The yearlings also enjoy tree shaking. But when they grow tired of active play they take up a quieter adult pastime—grooming. They have noticed that when a male wishes to be groomed by another male, he usually first mounts this male. The yearlings do not seem to understand that the mounting, which is being done by a superior to an inferior, is really a command to groom. They interpret it instead as a friendly offer to groom. So, when one of them mounts a fellow yearling, it lies down in the grooming posture. Then, gravely and earnestly, the other yearling practices on the small prostrate form.

Fall arrives, bringing with it the mating season. But the little monkey, secure in his yearling group, no longer feels the loneliness of separation, though he often loses sight of his mother for long periods at a time.

As the chill wintry gales return, however, his newfound sense of independence ebbs. Freezing temperatures usher in fierce blizzards, and the armies of the snow again invade the land. The yearling returns meekly to his mother. Once more riding high astride her back or clinging to her belly he travels tbrough the winter as safe as ever.

This will be the last year he can ride in this fashion. His mother is pregnant again, and when the new life arrives it will take his place and ride on his mother's back next year. The youngest in the Japanese monkey family is always the favored one because it stands most in need of the mother's protection.

Spring brings another birthing season. A tiny sister usurps the place of the former yearling, now a two-year-old juvenile. As a female, she will have much closer bonds with her mother than her brother ever had.

But the little juvenile is not concerned about this. As his adventurous spirit grows, the vistas before him open wider and wider. Now he and his small troop begin to move away from the center where the females and infants stay to venture into the periphery. For short periods, they linger to play with the three- and four-year-olds who have also strayed into the periphery, or with the young males, who like to tease them by grabbing them and mounting them.

Most of all, the little juveniles enjoy a frolic with the old solitary. They see no reason to shun him as the other monkeys do, for to their inexperienced eyes he is not much different from any of the other adults, except that he is more friendly, more tolerant of them.

Around the tattered, bedraggled old fellow they gather to romp. They leap at him, playfully cuffing him. He cuffs back gently. They nip him lightly and he nips back, all in good fun. They make false lunges at his throat. He lies on his back, his legs waving in the air, and lets them jump up and down on his belly. He sits on his haunches, and they scramble over his back and up his shoulders to perch on his head. He never scolds. But if one starts bullying another, he makes a threatening grimace that puts the bully in his place.

One day, in an exuberance of good feeling, the little juvenile mounts the shaggy rump of the former solitary. Then he begins to groom him. With a sigh of pleasure, the old male stretches out on the ground, looking like a worn-dirty old rug. Crouching over him, the juvenile begins earnestly to comb through the unkempt matted hair, parting it with deft strokes, picking, combing with his small fingers.

The solitary closes his eyes. A look of pure pleasure rests on his mottled, wrinkled old face. When has it been since he last had a grooming? He cannot remember.

# 9
# Death of a Leader

As the years slip by, the young juvenile visits the periphery more and more frequently. Presently he returns only occasionally to form a grooming party with his sister and mother, who has now born her second daughter.

Then, in the summer of his fifth year, he joins the company of the young males for good. Now he has two choices. He can work with the more responsible males—scouting, patrolling, doing sentry duty, urging on and rounding up laggards—or he can follow the course of the young rogue males. These males ignore all duties to satisfy their own whims, although they remain within the framework of the troop.

How appealing such a life must appear to the young juvenile! For the past five years he has lived under the strict disciplines of his mother, the other females, the leader, and the two subleaders. And so he shuns the call of duty. With two companions, he leaves the orderly trek to swing through the trees on adventures of his own.

*238* :

Sometimes he and his companions take shortcuts to arrive at the next resting place ahead of the others. Sometimes they roam deeper in the wilderness, joining up with the troop only at day's end. What a delight these summer vistas of unexplored forest are to the young monkeys.

Now and again they chance upon a solitary who is bigger and stronger than they are, for they have not yet attained their full growth. So, when the stronger monkey growls at them with his fierce grimace, they meekly turn around and present their buttocks to him. Only after he has mounted them one by one will he allow them to proceed on their way.

It is different under the protection of the troop. Here they can afford to act much more bravely. Whenever a solitary tries to enter the troop's periphery, they join with the other males to drive him away with triumphant squawks and barks.

In the spring of the young monkey's sixth year an uneasy atmosphere begins spreading through the troop. The old leader, the cause of this insecurity, is declining in health. His teeth, worn down by more than thirty years of service, have difficulty biting through hard substances. During the past winter he had to content himself with the meager sustenance he could get from twigs and small branches. Half starved on these scant rations, he almost died of pneumonia, which has left its mark in an overall weakness and fits of trembling.

Other signs of ill health and old age are also apparent. The leader's hair has become grizzled and rough instead of lying smoothly over his body. In his mottled face his eyes look out at times with a vague, unfocused stare. As the spring wears on, he begins to drag his hind legs. A slow paralysis is attacking his lower back. On the long foraging treks he hobbles painfully, stopping for frequent rest periods along the way. Despite his growing handicap, he continues to govern the troop. But now he must depend more and more on the subleaders to carry out his orders, which he indicates by a glare, a threat grimace, a low growl. As long as he is capable of controlling the troop, even though it is done by proxy, he will be accepted as the leader. After all, he has governed the troop well for more than fifteen years.

But by midsummer it is obvious his rule is almost over. Now he can no longer use his lower limbs and has to drag himself along by his hands. Still, every morning he rises to point out the day's trail. Obediently the troop follows it, quickly outdistancing him. Deserted by them, he drags himself over the ground painfully. Hours may pass before he finally overtakes them at some stopping place. Once there, he can depend on a restful grooming from the loyal females.

Other monkeys in the troop are becoming more and more inconsiderate, even insolent. Juveniles who before never dared cross his path chase one another freely right in front of his nose. Sometimes they even snatch a cluster of berries from his outstretched hand and eat it themselves.

Then, one day, a brash young male attempts to mount him. It is the final indignity. Slowly the old leader staggers to his feet, turns on the male, and barks *Ga! Ga!*

The male runs off, chased by the subleaders. But three days later the loyalty of the second subleader has broken. He, too, tries to mount the old male. Unable to defend himself any longer, the leader creeps away into the underbrush near the evening's campsite. Next morning, two juveniles discover his body there, stretched out in death.

A general wave of apathy settles over the troop. Instead of going on its daily foraging trek, it sits at the campsite in a stupor. First to rally is the highest-ranking female, the mother of the first subleader. Aged and practically toothless, she totters about the campsite settling the squabbles of the yearlings and juveniles and keeping them in order.

By the second day the old crone is pushed aside by the males. The apathy is over, and a rash of fighting breaks out. During the next three days it spreads throughout the troop, interspersed with sporadic foraging.

The female cliques are battling among themselves. The males are striving to raise their status with individual duals. Necks are wrung. Buttocks and bellies drip blood. One monkey staggers off with a dislocated shoulder. Another has lost an ear and a third an eye.

The bedraggled old solitary, once so submissive, has also changed in character. Perhaps some memory of his past as a high-ranking male is moving him with a new sense of duty and purpose. He rushes in to break up fights among the young males and to help keep order in the periphery.

At the top, where the two subleaders are battling for leadership, the struggle is fiercest. They chase each other around the campsite, snarling and clawing. Finally they disappear with loud roars into the woods beyond.

Now a solitary who has been lurking in the vicinity of the troop ever since the leader's death decides to make his move. He is a sinister-looking fellow, his body marked with the scars of many fierce battles. One finger has been broken and has set crookedly. One earlobe is missing and there is a long white slash along the edge of his mouth, pulling the corner down in a perpetual snarl.

He stalks toward the quarreling females and begins biting and wringing their necks, growling fiercely. Obviously he was a leader before he became a solitary, because he carries such authority that the females, shrieking and grimacing in fear, start running from him. Soon he has put an end to their squabbles. It is his first step in taking over the leadership.

But all at once a furious projectile of grizzled fur flings itself upon the solitary. It is the old mother of the first subleader. Shrieking at the top of her voice, she leaps on his back and clings to him, clawing and snapping at his neck with the few worn teeth she has left.

Her shrieks bring both subleaders rushing out of the woods. They charge the solitary just as he finally manages to shake himself free. But then a pack of young males, summoned by the commotion, rush in to the attack. The solitary is driven away, streaming with blood, his tail drooping, he disappears into the forest never to be seen by the troop again.

The two subleaders face each other, ready to continue their battle. Suddenly, the second subleader finds himself confronted by the fiery old female and other members of her clique. Knowing that he can never be leader unless the high-ranking females accept him, he acknowledges defeat by humbly turning his buttocks to the first subleader. This subleader mounts him triumphantly, proclaiming himself the acknowledged leader of the troop. The second subleader is promoted to the first subleader position. Now, forgetting their recent duals, the two work together to bring order out of chaos. Before the day is over, calm has settled on the troop.

Some monkeys have raised their positions as a result of the melee. Others have been demoted. Some of the demoted monkeys have defected to become solitaries or try to join another troop. They seldom return to their own.

The change in the bedraggled old solitary's appearance has become permanent. He has put on weight and his gaunt sides have filled out. His hair has become glossier, his eyes brighter. He has gained rank in the troop, and though it is low rank, his place is now secure.

A change has come over the young male too. Perhaps the disorders in the troop have wakened a sense of duty in him. Perhaps it is just that, at six years of age, he has reached sexual adulthood, though it will be at least two more years before he gains his full growth and the last of his milk teeth have been exchanged for permanent ones. That fall he mates for the first time, as does his four-and-a-half-year old sister, who has matured two years earlier than he.

By the time winter closes in, the young monkey has taken on sentry duty with other serious adult males. One freezing wintry day he is out scouting and spots several low, black forms loping over the hard-packed snowfields. As he stares, the forms become rapidly larger. They are coming his way, swiftly, silently. He recognizes them as one of the monkeys' most dreaded enemies, the feral dog, a domestic dog gone wild.

Instinct urges the young monkey to flee to the leafless woods where he can climb out of the reach of the pack. But his new sense of duty keeps him riveted to the spot. He sounds the warning cry, *Kuan! Kuan! Kuan!*

*Kuan! Kuan! Kuan!* It echoes again and again over the snowy slopes as it is relayed from monkey to monkey.

Everywhere, monkeys scatter for the trees. But the young male stands in place, still shrieking *Kuan!* still keeping watch on the loping dogs. He will stay there until he knows the leader has seen the danger and is taking charge.

Finally he hears the deep familiar voice roaring *Ga! Ga! Ga! Ga!* He is free at last to leave. Bounding over the snow, he makes for the nearest tree and shimmies up it, just ahead of the foaming pack. The dogs, crazed with winter famine, surge furiously around the trunk, yapping and howling.

In another tree the leader stands boldly on a branch in full sight of the dogs, growling *Ga! Ga! Ga! Ga!* He will not leave that perch until he is sure the dogs have gone for good. Only then will he give the signal to descend. This is his duty as troop guardian. But it is to the young male that the monkeys owe their lives, for without his vital warning, the dogs would have come upon them unaware.

# 10
# A New Branch

During the next two years, the young male takes on more and more responsibilities. In the beginning he made many mistakes. He would get so excited when a fight broke out that he would attack a nearby monkey instead of the culprit who started it. But he has since learned to control himself until he can reach the real offender.

His powerful fangs and muscular build have earned him so much respect that now he can usually end a fight with only a threat grimace and a *ga ga ga*. At eight years of age he has reached his full growth. He is close to three feet long and weighs more than forty pounds. He has risen to fifth-ranking place among the periphery males.

These past two years have been kind to the Alps. Winters have been milder. Heavy rainfalls coupled with days of brilliant sunshine have resulted in an abundance of edible flowers, foliage, fruits, berries, and nuts. The monkeys thrive on the rich food, growing fat and sleek.

Births have increased the population rapidly, more than doubling it.

Counting the newborn, there are now some seventy members. This would not be large as troops go in southern Japan. But in these Alps, with their extreme climate and limited food supplies, it is an unhealthy sign.

There are other danger signals. The new first subleader, who has a forceful nature, has never been satisfied with his subordinate role. One day he leaves to become a solitary. None of the capable young males are able to take his place because of the insurmountable barrier that seems to prevent them from crossing over into the center of power. Instead the position falls to a five-year-old juvenile whose high-ranking mother and her clique fight for him.

The new leader, unlike the old, is lax in discipline and leaves much of the policing of the troop to his brash young assistant, whom many of the females dislike. Quarrels break out, and some of the ordinary females start drifting to the periphery, where life is freer.

To make matters worse, the two years of plenty are followed by a lean year. A bitter winter ending in a drought causes a scarcity of food everywhere. Several infants die. Tempers become short among the adults. There are numerous fights.

One late afternoon of a chilly spring day the young male, who is now nearing his tenth birthday, lingers behind the main troop with a companion. Four females and their infants, none closely related to the young males, also stay behind to browze on the sparse spring foliage, while the main troop goes on to the night's campsite.

When the little group finally starts after the main troop, it is separated from it by almost a third of a mile. Instead of joining up at the campsite, it stops some distance away to spend the night by itself. It is the beginning of a split.

In the days that follow, the new branch of the troop lags farther and farther behind the main troop. By this time several more males have joined it, along with three curious juveniles who have defected to the new group in a body. Females, as though confused, pass easily from one group to the other, interchanging groomings. Finally, three more females stay with the new branch. They too, are not closely related to the males of this second group. It is as though the monkeys are instinctively trying to prevent inbreeding.

No matter how far apart the new group may be from the main troop during the daytime, it always sets up its night camp nearby. Here the monkeys are close enough to take part in the comfortable evening chatter of the mother troop. Back and forth, back and forth, they call to one another in a kind of antiphony that gives the branch a sense of security despite its separation.

As spring wears into summer, the sleeping places of the branch and the main troop spread farther and farther apart, until several ridges separate the two sites. The friendly sounds of the neighboring troop can no longer be heard.

Only occasionally now do the paths of the two troops cross. These chance encounters grow fewer and fewer as the new troop strives for complete independence. This is not easy because it will remain within the territorial limits set by the main troop. It must carve out new routes for itself in places that the main troop seldom visits. These places will naturally be poorer in food supplies. During the hot summer months, when provender is scarce anyway, the new troop will feel the want bitterly.

Other factors also work against its success. It has several more males than females. The balance has to tip the other way for the troop to survive. An excess of males causes them to fight fiercely among themselves. These attacks drive a number of males away, one after the other, setting the balance right. Now the new troop numbers twelve adults, seven female and five male.

Another danger is the loose social organization of these early days. At first there is no division of responsibility and no authority. Juveniles and adult males mix freely and without any regard to order. The females continue to wander around the periphery or stray into the forest, instead of forming a nucleus with the leader.

But gradually the young male begins to assert himself. Accepted by the females, he takes over the tasks of leadership, while the monkey who helped him start the new troop becomes subleader. Between them they administer demonstration attacks to prove their superiority. One by one they vanquish other males and mount them. They make control attacks on females who start brawls and on any monkey that does not show them the proper respect. In a short time threat grimaces alone are enough to keep the small troop in order.

By fall, fruits and berries and nuts spill their abundance over the mountains. The main troop forages far and wide to reap the autumn plenty. To avoid their former companions, members of the new troop have to confine themselves primarily to the northern sector of their territory. Food is scarcer here. The need for better foraging drives the new troop into an open stretch of country clothed only with tall grass, bushes, and an occasional tree. This open terrain is dangerous for the forest-dwelling monkey, whose chief defense is the shelter of trees.

Worse yet, the open land is also a domain of man, the monkeys' deadliest enemy. Looking down from the brow of the hill the little troop sees a village surrounded by fields of ripening soybeans. The sight of so much plenty lures the monkeys on. They are now in grave danger because, although the government has granted them protection, there is nothing to prevent an irate farmer from taking the law into his own hands and shooting the marauders.

Oblivious to this threat, the monkey scouts move forward eagerly. They see nothing in sight to harm them. But suddenly the leader, walking with the females, catches sight of a wire mesh limned against the sky, rising between the troop and the soybean fields. It is an electric fence, put up by the owner to take care of monkey predators.

The mesh is a strange object to the young leader. He has never seen its like before. To him it spells danger. Snarling and growling, he rushes forward, passing the scouts. He posts himself in front of the fence. Standing there he gives the warning cry, *Kuan! Kuan! Kuan!*

The subleader corrals the females and small monkeys into a tight group over which he stands guard. But several of the juveniles, goaded on by the sight of the laden bean bushes, continue to rush forward.

Their leader is ready for them. Here! There! Everywhere! He whirls around, nipping, cuffing, bowling them squawking head over heels. Finally they back away from him, away from the almost invisible fence and the rich bean harvest. Then, following his direction, the whole troop turns back.

Crying *Hui! Hui! Hui!* the monkeys make their way again to the safety of the forest cover. Leader and followers have proved themselves an integrated, well-disciplined unit, prepared to meet any danger that may come during the long, dreary winter months ahead.

# 11

# The Solitary

For three years the young leader vigorously governs his new troop. But as he enters his twelfth year a change comes over him. He begins to lose interest, leaving his subleader to take over more and more of his duties. Sometimes he wanders away by himself for two days at a time. It is as though the memories of his early, carefree days are drawing him back to the lonely forest again.

Gradually the young leader's status slips below that of his subleader, who now takes his place. For a while he acts as subleader, but his status continues to fall. Now the first male of the periphery has more power than he. Yet it is impossible for him, as former leader, to join the young males in the periphery. The barrier between periphery and center is just as difficult to cross downward, apparently, as it is to climb above it. There is no longer a place for him here.

One day he walks away from the troop for the last time. Now a solitary, he wanders alone through the green and golden days of late spring, his tail held proudly erect, for here in this wilderness his status remains unimpaired. He is king, a silent, shadowy king, making no sound because there is no one with whom to communicate.

Now and then in his wanderings he meets another solitary. The two quickly establish rank. The one who shakes a tree and roars more convincingly is the acknowledged victor, and the other presents his buttocks to be mounted. Then victor and vanquished part peaceably.

Once he is joined by a six-year-old monkey, and the two keep company for several days. During this time both become vocal. But then, wearying of each other's companionship, they separate.

Off goes the former leader, once more silent on his way. His travels take him far and wide, and he sees strange and wonderful things. One fall he joins a troop that in winter visits a high slope where strange two-legged creatures in bright warm costumes swoop down the snowy mountainside like birds.

The monkeys of this troop have learned that if they sit patiently at the foot of the ski slope, one of the creatures may take pity on them and give them handouts of apples or Japanese rice treats. Such handouts are not lavish because there are signs warning visitors against feeding monkeys.

The presence of so many boisterous human beings gathered in one place makes the solitary uneasy. Humans always spelled danger to his troop. He cannot adopt the nonchalant attitude of the other monkeys and soon drifts away into the wilderness to subsist on his usual homely winter fare of bark.

In the fall of another year he crosses several ridges to come upon a chattering river beside which a pool fed by hot springs casts up a faint mist of steam. A troop has made this its base because of the rich provender of grains, vegetables, and apples supplied by several two-legged creatures whom he watches warily for a while.

This is the Jigokudani Monkey Center, manned by Japanese scientists who feed the monkeys to bring them to this spot so that they can study their ways. The solitary waits behind some bushes on the upper slope. By this time he has grown a little more accustomed to human beings. So, after the other monkeys have finished eating and gone away, he gathers enough courage to come down and sample the food himself. He does not notice how the two-legged creatures carefully make a note of his appearance there.

He stays on through the fall and winter mating season, finding life much easier here than with his own little troop. There is always enough to eat. Even warmth is provided by the hot-water pool, although the solitary, along with the other males, does not take advantage of it. But, when snows and blizzards bring on freezing weather, juveniles and females carrying their babies slip into the pool and bask there, bodies warm and comfortable, snow wreathing their furry heads.

As the mating fever dies and the troop males and females become increasingly antagonistic toward him, the solitary wanders off. Does nature send him and other solitaries to wander and mate among the various troops on these mountains, mixing blood strains and preventing complete inbreeding? There is, after all, no wholesale intermating between the males and females of separate troops, which seldom if ever come in contact with one another.

Never does the solitary lose his pride. If a troop is so antagonistic that he may enter it only with lowered tail, he hovers in the vicinity long enough to issue a fiercely roared challenge and then passes on. If it is the mating season, he need only hover nearby to have some eager female search him out and even leave the troop to join him for a few days of courtship and mating.

So the years pass. At last he comes full circle to his old territory. One late afternoon, from high atop a tree he spies the troop he once led passing by, some on the ground, others swinging through the trees. Their chattering, squawks, cries, yaps, *kus*, and *huis* fill the sunset air with a familiar homely medley.

Something rises in him full and troubled. Instead of shaking the tree and giving his usual threat roar, he begins to call after them *kuuuu kuuuu kuuuu* in long, low tones.

There is no answer to his questing cry. The monkeys pass on. Their voices die away.

He calls again, the poignant *kuuuu* floating long and drawn out on the dying day.

Then he hears a female voice replying in soft, clear tones, *ku ku ku ku*.

Attentively he turns his head in that direction, hunched on the branch as if for action, his eyes alert, questioning. He waits and waits, while the colors fade from the sky and twilight gathers around him. But the *ku* comes no more. There is only the monotonous song of the nightjar, *errrrr*.

No longer is the solitary welcome in the troop which he founded. Silent, he sinks back into the cloak of falling night.

# Epilogue

The *Macaca fuscata* belongs to the order of Primates, itself a branch of the mammal family. About seventy million years ago, when the giant reptiles died out, the mammals began to diversify. One of these diversified branches was the primate branch.

At first the Primates lived in rock crevices, rotted logs, or tree roots, where they foraged for food. But by the middle Eocene epoch, some forty-five million years ago, they had become arboreal, with large front-facing eyes and short noses. Their legs became much longer than their arms, and their big toes separated from their other toes. This enabled them to grasp objects with their feet. They also evolved opposable thumbs so that they could hang on to slender branches.

It is not known how the *Macaca fuscata* arrived in Japan, but they probably came by a long vanished land bridge. Fossil remains show that these monkeys have inhabited the islands at least since the middle or late Pleistocene epoch, one to two million years ago. At that time the Japanese

climate was much warmer than it is today, and the islands were entirely covered with forests. The monkeys shared the area with the elephantlike Stegodon and the Palaeolonodon.

After human beings arrived on the islands, they became in time the monkeys' worst enemy. Hunters roamed the forests, first with primitive weapons and later with guns. Monkeys and other animals suffered at their hands.

In 1948 the Japanese government decided to step in because many rare animals were in danger of extinction. The Japanese wolf had already gone. The Japanese deer and wild boar had disappeared from Shiga Heights. Serow, raccoon dog, and badger had become rare. And the numbers of Asian black bears and unique macaques had diminished rapidly.

The Japanese government extended protection to many of these animals. It gave special recognition to the *Macaca fuscata,* designating it a national treasure and banning its export, capture, or killing. At this time its numbers had been reduced to several thousand.

The protection law was enacted in great part at the urging of Japanese scientists, who had begun to interest themselves in the fascinating *Macaca fuscata.* To study it, they founded the Kyoto University Primate Research Institute at Inuyama. Today the center is one of the largest in the world, dealing not only with the Japanese macaque but with the whole family of monkeys and apes.

To facilitate fieldwork with the monkeys, the scientists set up various stations in the national forest reserves, where monkey troops are situated. Here they enticed the monkeys with food to bring them into close range so that they could be studied more easily. As the monkeys gradually overcame their fear of humans, the scientists were able to learn more and more about their fascinating way of life. But, meanwhile, protection and overfeeding were causing the monkeys to increase rapidly until by 1962 some population estimates ran as high as 34,000 monkeys.

Increasing population brought fresh problems. Unable to find enough to eat in their traditional territories, the monkeys began overflowing into the farmers' fields and raiding their crops. Farmers sometimes took the law into their own hands and shot the marauding monkeys.

The scientists were now faced with a vicious circle. Providing the monkeys with more food in order to keep them out of the farmers' fields would only result in additional population explosions. The scientists decided that the best course was gradually to curtail the amounts doled out to the monkeys and to warn tourists against feeding them in the national reserves.

These actions have had little effect on the private monkey parks set up by enterprising businessmen. The attraction for tourists was that they could feed the animals as they pleased. This has resulted not only in an overpopulation of monkeys in more settled localities but also in alarming behavioral changes among the monkeys. Half tame, the monkeys have become overbearing, sometimes threatening, even invading the nearby tourist centers. The resultant ill will they have earned among shopkeepers brings with it a new threat of overkill.

By special permission, two havoc-causing troops were captured and shipped to the United States—one to Oregon, the other to Texas. Here they live in as near the wild state as possible, studied by American scientists. Other troops were captured and released in more extensive territories in Japan. But this can scarcely be considered a solution to the situation.

The crux of the problem is not overpopulation but the rapidly diminishing forestlands that are the monkeys' natural habitat. Spreading factory complexes, cities, towns, and farms all erode the Japanese wilderness. And there is a continuing outcry from business interests to release more and more pine-forested reserves that the government has designated as national parks. Even the Environmental Agency of the Japanese government is allowing itself to be swayed by industry's demands for more lumber, more roads, and more factories to bolster the economy.

Other problems are caused by new highways which local governments have been allowed to build through some of the reserves. These new highways are the scene of many accidents to monkeys and other wildlife. And the noise and pollution they cause disrupt the normal routine of the creatures that live on the reserves.

Another hazard is the aerial spraying with herbicides to keep down the undergrowth in whole sections of forest. Because the monkeys refuse to eat anything tainted with chemicals, they suffer real hardship, especially in the colder regions where winter fare consists entirely of bark and twigs.

A small but growing nucleus of environmentalists has come to the defense of the monkey and other wildlife. Their remonstrances have curtailed the use of monkeys in experimental laboratories in Japanese universities. And in 1978 the United States government was persuaded to ban the import of monkeys for such purposes.

The aim of Japanese environmentalists is a balanced ecology that would permit the survival of some of their country's loveliest woodlands and forest areas. Without those native forests, the Japanese macaque can survive only as a semidomesticated or zoo animal. This seems a sad anticlimax for a creature that for centuries has dominated Japanese folk legends—an endearing little trickster that usually outsmarts itself. It has also been accorded a certain minor divinity.

At Nara, an ancient Buddhist center and today a national park, an image of the monkey is enshrined in resplendent crimson robes. Visitors who come to Nara hoping to gain good luck, often visit the monkey shrine to pat the sly simian face of the little image.

It is the Japanese macaque that also serves as the model for the three famous monkeys of Buddhist lore that see no evil, hear no evil, speak no evil. Today's conservationists, desperately fighting for the rights of their nation's wildlife, say the monkeys are really representative of officials who "see no people, hear no people, speak to no people." The environmentalists are hoping that if they shout loudly enough, they may waken their fellow citizens to the value of the precious heritage that is irretrievably slipping away from them.

# Bibliography

PERIODICALS

Alexander, B.K., and Hughes, Jennifer. "Canine Teeth and Rank in Japanese Monkeys." *Primates* (Japan), March 1971.

Chang, S. "At Home in the Snow." *International Wild Life,* May–June 1978.

Eaton, G. Gray. "Social Order of Macaques." *Scientific American,* October 1976.

Furuya, Yoshio. "An Example of Fission of a Natural Troop of Japanese Monkeys at Gagyusan." *Primates* (Japan), March 1962.

———."On the Fission of Troops of Japanese Monkeys II—General View of Troop Fission of Japanese Monkeys." *Primates* (Japan), March 1969.

———. "On the Malformation Occurred in the Gagyusan Troop of Wild Japanese Monkeys." *Primates* (Japan), December 1966.

Gouzoules, Harold: Fedigan, Larry; Fedigan, Linda M. "Responses of a Transplanted Troop of Japanese Macaques *(Macaca fuscata)* to Bobcat *(Lynx rufus)* Predation." *Primates* (Japan), September 1975.

Hahn, Emily. "A Reporter at Large: Days at the Zoo." *New Yorker,* September 1967.

Hayashi, Katsuji. "Utilization of Ledges by Japanese Monkeys in Hakusan National Park." *Primates* (Japan), June 1969.

Hazama, Naonosuke. "Weighing Wild Japanese Monkeys in Arashiyama." *Primates* (Japan), 1964.

Ikeda, Jiro, and Watanabe, Tsuyoshi. "Morphological Studies of *Macaca fuscata* III Craniometry." *Primates* (Japan), June 1966.

Itani, Junichiro. "Paternal Care in the Wild Japanese Monkey *Macaca fuscata. fuscata.*" *Primates* (Japan), March 1962.

Iwamoto, Mitsuo. "Morphological Studies of *Macaca fuscata.*" *Primates* (Japan), June 1971.

————. "On a Skull of a Fossil Macaque from the Shikimizu Limestone Quarry in the Shikoku District, Japan." *Primates* (Japan), March 1975.

————, and Hasegawa, Hoshikazu. "Two Macaque Fossil Teeth from the Japanese Pleistocene." *Primates* (Japan), March 1972.

Izawa, Kosei. "Japanese Monkeys Living in the Okoppe Basin of the Shimokita Penninsula: The First Report of the Winter Follow-up Survey after the Aerial Spraying of Herbicide." *Primates* (Japan), June 1971.

————. "Japanese Monkeys Living in the Okoppe Basin of the Shimokita Penninsula: The Second Report of the Winter Follow-up Survey after the Aerial Spraying of Herbicide." *Primates* (Japan), June 1972.

Kawai, Masao. "A Case of Unseasonable Birth in Japanese Monkeys." *Primates* (Japan), September 1966.

"A Field Experiment on the Process of Group Formation in the Japanese Monkey (*Macaca fuscata*) and the Releasing of the Group at Ohirayama." *Primates* (Japan), March 1962.

————. "Newly Acquired Precultural Behavior of the Natural Troop of Japanese Monkeys on Koshima Island." *Primates* (Japan), 1965.

Kawamura, Syunzo. "The Process of Subculture Propagation among Japanese Macaques." *Primates* (Japan), March 1962.

Kawanaka, Kenji. "Intertroop Relationships Among Japanese Monkeys." *Primates* (Japan), September 1973.

*Life* "Snowbound Monkeys in a Hot Bath" (N.Y.) January, 30, 1970.

Lucas, Christopher. "The Marvelous Monkeys of Japan." *Readers Digest*, November 1968.

Miyadi, Denzaburo. "Social Life of Japanese Monkeys." *Science*, February 1964.

Mizuhara, Hiroki. "Social Changes of Japanese Monkey Troops in the Takasakiyama." *Primates* (Japan), 1964.

Mori, Akio. "Intra-troop Spacing Mechanism of the Wild Japanese Monkeys of the Koshima Troop." *Primates* (Japan), April 1977.

———. "Signals Found in the Grooming Interactions of Wild Japanese Monkeys of the Koshima Troop." *Primates* (Japan), June 1975.

Murray, R. Daniel, and Murdoch, Katherine Mayer. "Mother-Infant Behavior in Japanese Monkeys." *Primates* (Japan), October 1977.

Nigi, Hideo. "Laperoscopic Observations of Ovaries Before and After Ovulation in the Japanese Monkey *(Macaca fuscata)*." *Primates* (Japan), April 1977.

———. "Menstrual Cycle and Some other Related Aspects of Japanese Monkeys *(Macaca fuscata)*." *Primates* (Japan), June 1975.

———. "Some Aspects Related to Conception of the Japanese Monkey." *Primates* (Japan), January 1976.

Nishida, Toshisada. "A Sociological Study of Solitary Male Monkeys." *Primates* (Japan), June 1966.

Norikoshi, Kohshi. "Tests to Determine the Responsiveness of Free-Ranging Japanese Monkeys in Food-Getting Situations." *Primates* (Japan), June 1971.

Saheki, Masatomo. "Morphological Studies of *Macaca fuscata* IV Dentition." *Primates* (Japan), December 1966.

Sugiyama, Yukimaru. "On the Division of a Natural Troop of Japanese Monkeys at Takasakiyama." *Primates* (Japan), March 1962.

Suzuki, Akira. "An Ecological Study of Wild Japanese Monkeys in Snowy Areas Focused on Their Food Habits." *Primates* (Japan), August 1965.

———. "On the Problems of the Conservation of the Japanese Monkey on the Boso Penninsula, Japan." *Primates* (Japan), September 1972.

Takeda, Rumiko. "Development of Vocal Communication in Man-Raised Japanese Monkeys." *Primates* (Japan), December 1965.

Takuda, Kisaburo. "Sexual Behavior in the Japanese Monkey Troop." *Primates* (Japan), 1961–62.

Tasumi, Motoo. "Senile Features in the Skeleton of an Aged Japanese Monkey." *Primates* (Japan), December 1969.

Tokura, Hiromi. "Sweating in the Japanese Macaque." *Primates* (Japan), March 1975.

Yamada, Munemi. "Five Natural Troops of Japanese Monkeys in Shodoshima Island (I)." *Primates* (Japan), September 1966.

———. "Five Natural Troops of Japanese Monkeys on Shodoshima Island: A Comparison of Social Structure." *Primates* (Japan), June 1971.

Yotsumoto, Nobuko. "The Daily Activity Rhythm in a Troop of Wild Japanese Monkeys." *Primates* (Japan), April 1976.

## BOOKS

*Area Handbook for Japan*. Washington, D.C., 1969.

Bourne, Geoffrey H. *Primate Odyssey*. New York, 1974.

Carpenter, C. R., ed. *Behavioral Regulators of Behavior in Primates*. Pennsylvania, 1973.

DeVore, Irven. *Primate Behavior*. New York, 1965.

Freeman, Dan. *The Love of Monkeys and Apes*. Hong Kong, 1977.

*Grzimek's Animal Life Encyclopedia*, vols. 7–13. New York, 1972.

Imanishi, Kinji, ed. *Japanese Monkeys: A collection of Translations*, Japan, 1965.

Isida, Ryuziro. *Geography of Japan*. Tokyo, 1961.

Johnson, Hugh. *International Book of Trees*. New York, 1973.

Jolly, Alison. *The Evolution of Primate Behavior*. New York, 1972.

*Larousse Encyclopedia of Animal Life*. New York, 1967.

Napier, Prue. *Monkeys and Apes*. New York, 1972.

Novak, Frantisek A. *Pictorial Encyclopedia of Plants and Flowers*. New York, 1966.

Rosen, S. I. *Introduction to the Primates*. New Jersey, 1974.

Rosenblum, Leonard A., ed. *Primate Behavior,* vol. 2. New York, 1971. Includes the following article about the snow monkey:
Brandt, E. M., and Mitchell, G. "Parturition in Primates: Behavior Related to Birth."
———. *Primate Behavior,* vol. 4. New York, 1975. Includes the following articles about the snow monkey:
Green, Steven. "Variation of Vocal Pattern with Social Situation in the Japanese Monkey *(Macaca fuscata):* A Field Study."
Sanderson, Ivan T. *Monkey Kingdom.* Pennsylvania, 1957.
Simonds, Paul E. *The Social Primates.* New York, 1974.
Time-Life Films. *Monkeys and Apes.* New York, 1976.
Trewartha, G. W. *Japan.* Wisconsin, 1945.
Walker, E. P. *Mammals of the World,* vols. 1 and 2. Maryland, 1968.
Wit, H. C. D. *Plants of the World.* Translated by A. G. Pomerans. 3 vols. New York, 1966.


## BOOKS TRANSLATED FROM THE JAPANESE

Akihiko, Maezawa. *Mountain Botany.* Tokyo, 1970.
*Compilation of Plants in Shiga Heights.* Shinshu, 1962.
Kobayashi, Keisuke. *Common Birds of Japan in Color.* Tokyo, 1967.
Masui, Mitsuko. *The Animals of Japan.* Translated by Yeiko Nakashima. Tokyo, 1978.
Nakamura, Noboru. *Guide to Wild Birds.* Translated by Aki Tsurukame. Tokyo, 1977.
*Nature Research Papers of Arashiyama,* nos. 5 and 6. Tokyo, 1977.

## PERSONAL INTERVIEWS

Tokui Aiba, Radio Japan.

Shigeru Azuma, Japan Monkey Center, Inuyama.

Edward E. Hain, Radio Japan.

Ishikawa, Kazuhiko, producer, television documentary, *"The Snow Monkeys of Japan,"* Japan, 1977.

Shigetaka Kotera, Primates Zoo Japan Monkey Center, Inuyama.

Dr. Naoki Koyama, Japan Monkey Center, Inuyama.

Satsue Mito, Koshima laboratory.

Dr. Akio Mori, Primate Institute of Kyoto University, Koshima.

Dr. Noboru Nakamura, Institute of Natural Education in Shiga Heights.

Chikao Nakata, Arashiyama Monkey Center.

Dr. Yukimaru Sugiyama, Japan Monkey Center, Inuyama.

Nobuo Suzuki, University of Hokkaido.

Eishi Tokita, Jigokudani Monkey Center.

www.ingramcontent.com/pod-product-compliance
Lightning Source LLC
Chambersburg PA
CBHW061342280526
45784CB00001B/94